CHEMISTRY RESEARCH AND APPLICATIONS

OXALATE

STRUCTURE, FUNCTIONS AND OCCURRENCE

CHEMISTRY RESEARCH AND APPLICATIONS

Additional books and e-books in this series can be found on Nova's website under the Series tab.

CHEMISTRY RESEARCH AND APPLICATIONS

OXALATE

STRUCTURE, FUNCTIONS AND OCCURRENCE

ELSA KYTÖNEN
EDITOR

Copyright © 2020 by Nova Science Publishers, Inc.

All rights reserved. No part of this book may be reproduced, stored in a retrieval system or transmitted in any form or by any means: electronic, electrostatic, magnetic, tape, mechanical photocopying, recording or otherwise without the written permission of the Publisher.

We have partnered with Copyright Clearance Center to make it easy for you to obtain permissions to reuse content from this publication. Simply navigate to this publication's page on Nova's website and locate the "Get Permission" button below the title description. This button is linked directly to the title's permission page on copyright.com. Alternatively, you can visit copyright.com and search by title, ISBN, or ISSN.

For further questions about using the service on copyright.com, please contact:
Copyright Clearance Center
Phone: +1-(978) 750-8400   Fax: +1-(978) 750-4470   E-mail: info@copyright.com.

NOTICE TO THE READER

The Publisher has taken reasonable care in the preparation of this book, but makes no expressed or implied warranty of any kind and assumes no responsibility for any errors or omissions. No liability is assumed for incidental or consequential damages in connection with or arising out of information contained in this book. The Publisher shall not be liable for any special, consequential, or exemplary damages resulting, in whole or in part, from the readers' use of, or reliance upon, this material. Any parts of this book based on government reports are so indicated and copyright is claimed for those parts to the extent applicable to compilations of such works.

Independent verification should be sought for any data, advice or recommendations contained in this book. In addition, no responsibility is assumed by the Publisher for any injury and/or damage to persons or property arising from any methods, products, instructions, ideas or otherwise contained in this publication.

This publication is designed to provide accurate and authoritative information with regard to the subject matter covered herein. It is sold with the clear understanding that the Publisher is not engaged in rendering legal or any other professional services. If legal or any other expert assistance is required, the services of a competent person should be sought. FROM A DECLARATION OF PARTICIPANTS JOINTLY ADOPTED BY A COMMITTEE OF THE AMERICAN BAR ASSOCIATION AND A COMMITTEE OF PUBLISHERS.

Additional color graphics may be available in the e-book version of this book.

Library of Congress Cataloging-in-Publication Data

Names: Kytönen, Elsa, editor.
Title: Oxalate: structure, functions and occurrence / Elsa Kytönen.
Identifiers: LCCN 2020030179 (print) | LCCN 2020030180 (ebook) | ISBN
  9781536183030 (paperback) | ISBN 9781536183528 (adobe pdf)
Subjects: LCSH: Oxalates.
Classification: LCC QD305.A2 O93 2020 (print) | LCC QD305.A2 (ebook) | DDC 547/.437--dc23
LC record available at https://lccn.loc.gov/2020030179
LC ebook record available at https://lccn.loc.gov/2020030180

Published by Nova Science Publishers, Inc. † New York

CONTENTS

Preface   vii

Chapter 1   Pathophysiologic and Therapeutic Aspects of Renal Calculi   **1**
Chitrakshi Consul and Sonu Chand Thakur

Chapter 2   Transition Metal Oxalates as Potential Futuristic Materials for Efficient Energy Storage Capacity   **55**
Sushree Pattnaik, Arya Das, Kishor Kumar Sahu, Suddhasatwa Basu and Mamata Mohapatra

Chapter 3   Oxalate Biosynthesis and Degradation in Plants and Fungi   **95**
Enrique J. Baran

Chapter 4   The Association of Diabetes and Obesity with Urolithiasis: A Review   **133**
Mohd Sajad and Sonu Chand Thakur

Index   **151**

PREFACE

Oxalate: Structure, Functions and Occurrence first summarizes the various factors contributing to calculi formation, along with different therapeutic measures including medicinal plants/drugs which have been experimentally proven to exhibit antiurolithiatic activity employing various mechanisms.

The authors discuss the known transition metal oxalates, their synthetic analogues, thermal and magnetic behaviour, and reaction pathways, particularly for energy materials such as Li ion, as well as supercapacitor and redox flow battery systems.

A number of possible synthetic pathways for the generation of oxalic acid in plants and fungi are presented and briefly discussed. An important number of studies have shown that ascorbic acid is the major substrate for the synthesis of oxalic acid in plants. Therefore, the so called "Wheeler-Smirnoff" mechanism, which explains the biosynthesis of ascorbic acid in the plant kingdom, is discussed in detail.

Lastly, this compilation explores how hyperoxaluria can be a pathway through which kidney dysfunction occurs in persons due to diabetes mellitus or obesity and thus lead to a gradual deterioration of renal function.

Chapter 1 - The term urolithiasis can be understood as the condition of formation of calculi (stones) that may take place in any part of the urinary tract. According to various studies conducted so far, urolithiasis may be regarded as one of the oldest diseased conditions affecting mankind since ancient times. The consequences of the same may include renal damage along with stone recurrence in some cases. Formation of stones in the kidney is a multi step process which includes nucleation, crystal growth, crystal aggregation and crystal retention. There are various therapies available for the treatment of kidney stones like painkillers, allopurinol, diuretic, computed tomography dietary modifications, ureterenooscopic and nephrolithotomy. The treatment for the same employs use of surgical techniques also. However these treatments have several side effects and there are chances of its reoccurrence. So there is a need for better and safer options. Several medicinal plants and herbal drugs have been reported to exhibit therapeutic efficacy to treat urolithiasis. Different drugs derived either from a single or combinations of medicinal plants have been proven to either dissolve or prevent calculi formation. So these medicinal plants could be a source for the alternative treatment of this disease. The studies in both humans and animal models of hyperoxaluria-associated calcium oxalate kidney stone disease have displayed effective results against urolithiasis without the risk of side effects and there is a prominent reduction in hyperoxaluria followed by diminished rate of recurrent stone formation. The content summarizes various factors contributing to calculi formation along with different therapeutic measures including medicinal plants/drugs which are experimentally proven to exhibit antiurolithiatic activity employing various mechanisms.

Chapter 2 - Transition Metal oxalate is one of the pragmaticmaterial forcatalytic, sensor, battery and SCs application due to its low cost, environmental friendliness, and good chemical stability. In this regard a lot of research work has recently focused on implementation of it as potential electrode materials for enhanced energy storage. Owing to their unique property to act as a carbon sink further establishes them as sustainable and green materials for energy storage, thus strengthening their stand among variety of materials available for possible commercialization. This progress report presents the various known transition metal oxalates, their synthetic analogues, thermal and magnetic behaviour, the reaction pathways, particularly for energy materials such as Li ion, supercapacitor and redox flow battery systems. Finally, the challenges of these materials in these energy stoarge systems for high power and energy output are outlined.

Chapter 3 - A number of different possible synthetic pathways for the generation of oxalic acid in plants and fungi are presented and briefly discussed. An important number of studies have shown that ascorbic acid is the major substrate for the synthesis of oxalic acid in plants. Therefore, the so called "Wheeler-Smirnoff" mechanism, which explains the biosynthesis of ascorbic acid in the plant kingdom, is discussed in detail. The possible involvement of D-erythroascorbic acid in the generation of fungal $H_2C_2O_4$ is also analyzed. The general characteristics of the two major oxalate degradation enzymes, oxalate oxidase and oxalate decarboxylase, are thoroughly discussed. Both systems are Mn(II)-depending enzymes, associated with the "cupin" super family of proteins, and their reaction mechanisms are closely related. Finally, the role of oxalotrophic bacteria in the oxalate-degradation processes is also briefly commented.

Chapter 4 - Hyperoxaluria can induce kidney disease by several pathways, including calcium oxalate crystal tubular blocking, inflammation, and injury in epithelial tubular cell. Hyperoxaluria is also seen in people with obesity and mellitus, which are extenuating features that lead to chronic kidney disease (CKD). In addition. This is not still clear that hyperoxaluria is a possible cause behind this increased incidence of CKD in diabetes mellitus and obesity. These are linked with higher excretion of urinary oxalate through a variety of pathways. Hyperoxaluria can be a pathway through which the kidney dysfunction occurs in persons due to diabetes mellitus or obesity and thus, lead to a gradual deterioration of renal function. Potential work on pharmacological or dietary interventions to minimize the production or absorption of oxalate is required to check whether the reducing excretion of urinary oxalate is helpful in inhibiting the development of kidney disease and the process of obesity and diabetes mellitus.

In: Oxalate
Editor: Elsa Kytönen

ISBN: 978-1-53618-303-0
© 2020 Nova Science Publishers, Inc.

Chapter 1

PATHOPHYSIOLOGIC AND THERAPEUTIC ASPECTS OF RENAL CALCULI

*Chitrakshi Consul and Sonu Chand Thakur**

Centre for Interdisciplinary Research in Basic Sciences,
Jamia Millia Islamia, New Delhi, India

ABSTRACT

The term urolithiasis can be understood as the condition of formation of calculi (stones) that may take place in any part of the urinary tract. According to various studies conducted so far, urolithiasis may be regarded as one of the oldest diseased conditions affecting mankind since ancient times. The consequences of the same may include renal damage along with stone recurrence in some cases. Formation of stones in the kidney is a multi step process which includes nucleation, crystal growth, crystal aggregation and crystal retention. There are various therapies available for the treatment of kidney stones like painkillers, allopurinol, diuretic, computed tomography dietary modifications, ureterenooscopic and nephrolithotomy. The treatment for the same employs use of surgical techniques also. However these treatments have several side effects and there are chances of its reoccurrence. So there is a need for better and

---

* Corresponding Author's Email: sthakur@jmi.ac.in.

safer options. Several medicinal plants and herbal drugs have been reported to exhibit therapeutic efficacy to treat urolithiasis. Different drugs derived either from a single or combinations of medicinal plants have been proven to either dissolve or prevent calculi formation. So these medicinal plants could be a source for the alternative treatment of this disease. The studies in both humans and animal models of hyperoxaluria-associated calcium oxalate kidney stone disease have displayed effective results against urolithiasis without the risk of side effects and there is a prominent reduction in hyperoxaluria followed by diminished rate of recurrent stone formation. The content summarizes various factors contributing to calculi formation along with different therapeutic measures including medicinal plants/drugs which are experimentally proven to exhibit antiurolithiatic activity employing various mechanisms.

1. INTRODUCTION

Urolithiasis is a diseased condition of formation of crystalline mineral deposits (stones) anywhere in the urinary tract [1].Urolithiasis can be understood as a combined term for Nephrolithiasis (stone formation in kidney; renal lithiasis), ureter lithiasis (stone formation in the ureters), Cystolithiasis (bladder lithiasis) and urethral lithiasis (stone formation in the urethra)[2]. It can be regarded as one of the oldest known diseases. Archaeological evidences after examination of Egyptian mummies reveal that people suffer from kidney stone diseases since ancient times [3]. Approximately, 12% of world population is estimated to be affected with kidney stones [1]. The most common problem associated with urolithiasis is the prevention of renal stone recurrence. Kidney stones are found to contribute to and increase the risk of a number of serious disorders such as chronic kidney diseases [4] cardiovascular diseases [4], end-stage renal failure [2, 4] diabetes and hypertension [4]. The symptoms associated with the same may be hematuria (blood in urine), pain in abdomen, flank (pain in the back side) or groin areas, obstructive uropathy (urinary tract disease), blockage of urinary flow, hydronephrosis (dilation of the kidney), decreased urine volume, increase in stone forming compounds and other of signs of infection [4]. Hence, urolithiasis can be regarded as a multifactorial disorder resulting from pathological biominerlization [5].

Drugs from some plants are designed and are reported to have anti-urolithiatic activity. Anti-urolithiatic drugs are those which dissolve or prevent urinary calculi formation. This involves a number of mechanisms such as maintaining crystalloid-colloid balance by decreasing the excretion of stone forming elements such as calcium, oxalate, uric acid, phosphorus, and proteins; contributing to increased excretion of urea and keratinine thereby, improving renal function, these drugs have potential antioxidant activity and also reduces renal tubular epithelial cell injury [6].

2. EPIDEMIOLOGY OF KIDNEY STONE

The rate of prevalence of urolithiasis was found to be 5-15% in low to middle income countries while it was found to be 1-5% in high income countries [7]. At low socioeconomic levels high frequency of urolithiasis is observed in children i.e., about 40%, while it was found to be nearly 20% in females. At high socioeconomic levels, highest frequency of kidney stones in observed among adults i.e., around 25% in females. Among both the groups the frequency of calcium oxalate stones was observed to be maximum that is about 40% and 60% respectively while the second highest frequency observed was that of uric acid/urate stones estimated to be around 30% and 40% respectively [6]. In context to Asia, the prevalence of urolithiasis was found to be 5-19% in West Asia, Southeast Asia, South Asia and some developed countries such as South Korea and Japan; while it was estimated to be about 1-8% in most parts of East and North Asia. In India and Malaysia, the incidence of urolithiasis was lower than 40/100000 while a sharp increase to about 930/100000 and 442.7/100000 was recorded 3 decades later [7]. However, the lifetime reoccurrence rate in males was perceived to be higher while the frequencies of nephrolithiasis seem to increase in females. A drastic increase in reported cases of urolithiasis in Europe, North America, South America, Australia and other parts of the world [4]. As per the statistics observed with regard to urolithiasis, it is found to increase over a few decades. There are certain other factors that are said to influence the frequency of urolithiasis like lifestyle modifications such as lack of physical activity, change in dietary habits, global warming etc.

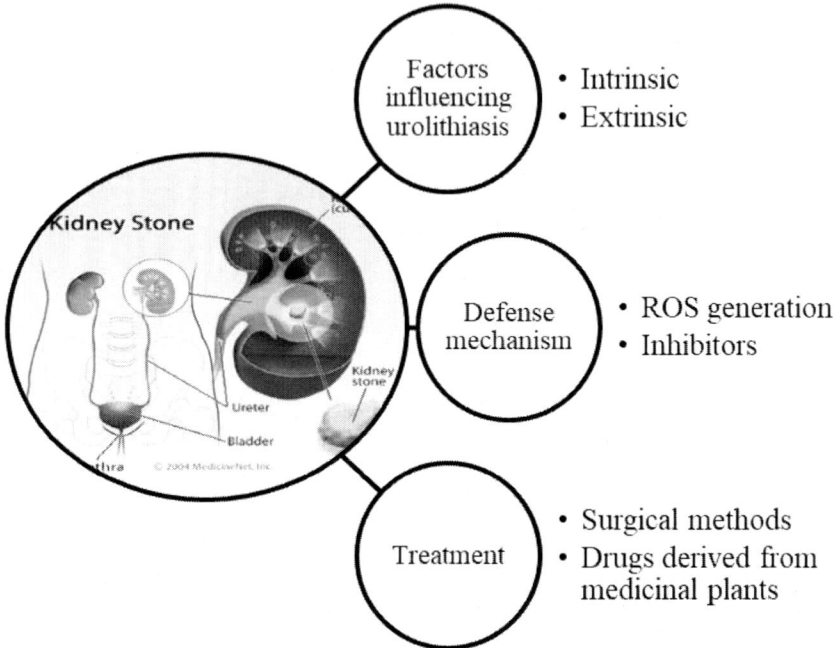

Figure 1. A basic overview of urolithiasis.

3. FACTORS INFLUENCING UROLITHIASIS

Factors influencing the prevalence of urolithiasis may be categorised as Intrinsic- including age, gender, race, ethnic and familial backgrounds; and Extrinsic factors such as lifestyle and dietary habits, climate and environment, occupational and educational level.

3.1. Age

Among the various factors listed, age forms one of the major factors for prevalence of urolithiasis. It is estimated that the incidence of urolithiasis is highest within middle aged population i.e., 30-60 years which is however, observed to decrease after this period. The reason behind this may be more laborious work thereby resulting in less fluid intake and high rate of dehydration. Also, males are more likely to be affected at the age of 70 years while the females at the age of 50 years [8].

3.2. Gender

Gender is another factor which is said to influence the prevalence of urolithiasis. The estimated ratio among males and females for the prevalence of urolithiasis is 1.3 to 5. Hormones such as testosterone are observed to promote urolithiasis while estrogen is said to inhibit the stone formation by regulating the synthesis of 1, 25-dihydroxy-vitamin-D. Anatomical differences such as benign prostrate hyperplasia in males, leading to obstruction of urethra is also considered a risk factor. However, it is observed that the ratio has declined within Asian countries which indicate that females seem to be at a higher risk of getting affected with urolithiasis [8].

3.3. Genetics and Obesity

According to the latest statistics, Genetics and Obesity were also evaluated as factors contributing to kidney stone formation. Genetic factors such as mutation in the SLC3A1, SLC7A9 gene or other neighbouring ones are said to influence Cystine stone formation. However, mutation in SLC22A12 and SLC2A9 were observed to influence uric acid stone formation. Stones composed of Calcium oxalate were found to be formed due to deficiency of enzymes like alanine glyoxalate aminotransferase (AGXT), glyoxlate reductase/hydroxypyruvate reductase (GRHPR) or 4-hydroxy-2-oxoglutarrate aldolase (HOGA1) while obesity is currently a known risk factor for stone formation [1, 8].

3.4. Race

Race is yet another factor responsible for prevalence of urolithiasis. The prevalence and incidence rates were observed to be highest in whites followed by Hispanics, blacks and then the Asians. Among these, white men were found to have the highest incidence of kidney stones while Asian women were observed to have the lowest rate [1].

Considering the extrinsic factors, lifestyle and dietary habits form one of the crucial factors influencing urolithiasis.

3.5. Diets

Diets including consumption of cereals and vegetables having high content of oxalate and its precursors like glycine, hydroxyproline, hydroxyl acetic acid and vitamin-C are said to influence urolithiasis. Also, westernized diets containing excessive protein, lipid, calcium and sodium are also said to contribute to kidney stone formation especially in Asian countries [8]. The two most important conditions for kidney stone formation i.e., pH and ingredient concentration should be fulfilled. In areas having rice as a principle crop such as South and Southwest Asia, are also prone to urolithiasis as the carbohydrate present in rice gets catabolised to provide acidic environment. Less fluid intake along with high levels fluoride, sodium, calcium, magnesium and phosphates in drinking water also form risk factors for calcium oxalate stone formation. The mechanism underlying this issue may be that presence of fluoride inside the intestine indirectly promotes oxalate absorption due to calcium fluoride formation leading to decrease in calcium availability [8].

Increased oxalate excretion and formation of insoluble calcium fluoride in urine along with oxidative stress in renal system are the resulting consequences. Apart from fluoride, presence of sodium is also regarded as a significant factor for its contribution to urolithiasis. Increased sodium levels lead to extra absorption of calcium into blood or inhibit absorption of calcium from urine into renal tubular epithelial cells, this is

Pathophysiologic and Therapeutic Aspects of Renal Calculi 7

intern is followed by enhanced excretion and precipitation of calcium in kidney. A theory contrasting this fact says, high sodium concentration is said to inhibit citrate excretion in urine; a crucial step in preventing crystal formation. High magnesium concentration in soil and drinking water may also contribute to stone development. Other habits such as smoking also form risk factors contributing to stone formation. Other than this, inadequate physical activity is also considered to contribute to stone formation. Therefore, people having sedentary lifestyle in office are said to be more prone to urolithiasis. Deficiency of certain microelements such as molybdenum and silicon found to play a key role in keeping crystal solution also contribute to stone formation [8].

3.6. Climatic and Environmental Factors

Climatic and environmental factors such as factors such as temperature, sunshine hours, seasons, humidity, rainfall and atmospheric pressure are also found to contribute to frequency of urolithiasis. Higher prevalence of urolithiasis is observed in regions of tropical and subtropical areas (5%-10%) as compared to temperate and frigid zones (1%-5%). Hot-dry climate results in evaporation of body water from the skin leading to urine concentration causing precipitation of crystals and stone formation. Such climatic conditions are usually observed in West Asia. In areas such as East China, the prevalence of urolithiasis is observed to hike during summer and autumn seasons which tend to reduce during spring and winter seasons. Other countries such as India, Pakistan, Saudi Arabia and Iran are observed to portray similar statistics. The reason behind this may be related to temperature threshold value. The rate of prevalence may be higher in areas having temperature range similar to the threshold (i.e., 18°C in Korea), as compared to regions having lower or higher temperature ranges. Occupation and education level are also said to influence the prevalence of urolithiasis. In countries such as Malaysia and Thailand, people working outdoors or those exposed to high temperatures such as workers in steel industry, flight attendants, farmers, miners, quarrymen or drivers were twice more prone to urolithiasis. Therefore, frequency of urolithiasis in high temperature countries was found to be high; the underlying reason may be dehydration caused due to high temperature and less access to drinking water. It is also observed that, excessive exposure to sunlight results in excess production of vitamin-D. This intern gets converted to 1, 25-dihydroxy vitamin D in the kidney thereby, promoting absorption of calcium in the gut. On the contrary, prevalence of urolithiasis was observed to be high in people working in offices or similar areas especially in countries like Japan, Saudi Arabia and Iran. The possible reason behind such findings may be the sedentary lifestyle including lack of outdoor activities and reduced fluid intake. Other factors responsible for urolithiasis include decreased urine output, hyperoxalurea (excessive urinary excretion of oxalate), hypercalciuria (excessive urinary calcium), hyperuricosuria (excessive urinary excretion of uric acid) and hypocitraturia (decreased concentration of citrate in urine) [8].

4. Mechanism of Urine Formation and Factors and Organs Associated with It

The mechanism of formation of urine takes place in the excretory system of the human body. The nephron present inside the kidney consists of a tubule that is closed at one end forming the Glomerular capsule while the other end opening into the collecting tubule. The part of the nephron enclosed between the glomrulus and the collecting tubule include; Proximal convulated tubule (site for solute reabsorption especially for chloride, glucose, sodium, and water), Medullary loop (loop of Henle) (site for urine concentration composed of 2.5% urea, 2.5% mineral mixture, salts, hormones and enzymes apart from 95% water), Distal convulated tubule leading to the collecting duct (site for minor adjustments in the composition of urine).

4.1. Filtration

The process of glomerular filtration takes place through the semipermiable walls of the glomrulus and glomerular capsule. This process allows separation of larger substances such as Blood cells, plasma proteins, etc., from smaller ones such as water and other smaller constituents. Therefore, the process of filtration takes place to balance the blood pressure between the glomrulus and the glomerular capsule.

4.2. Selective Reabsorption

Most of the filtrate generated during the process of filtration is reabsorbed by the different parts of the nephron. The constituents of the filtrate may be absorbed actively such as water, electrolyte, and organic nutrients including glucose etc., while others are passively reabsorbed. Components such as sodium and chloride can be absorbed both actively as well passively. Absorption of nitrogenous bases such urea, uric acid and creatinine is limited. Excessive absorption or restricted secretion of components such as urea, calcium, oxalate, cystine, phosphates, straitin and uric acid contribute to the formation of urinary calculi. However, composition of such calculi depends upon the non secreted components. [4, 35, 68].

4.3. Tubular Secretion

The peritubular capillaries clear the filtrate and drain it into the convulated tubule. The composition of tubular secretion includes the components that are partially filtered during the process of filtration. This includes certain drugs such as penicillin and aspirin, hydrogen ions (important for maintaining blood pH).

The urine composition is the prima facie indicator of substances exchanged between nephron and blood present in renal capillaries. Waste products obtained from protein metabolism are excreted; pH (acid-base balance) is maintained by hydrogen ion excretion, water and electrolyte levels are controlled [4, 68].

5. KIDNEY STONE – TYPES AND COMPOSITION

The type of kidney stone is determined by the abnormal constituent that skips excretion. Such stones differ in size, chemical composition and shape. Kidney stones on the basis of pathogenesis and mineral composition are classified into four basic types among which the most prevalent form of urinary calculi being the Calcium stones.

5.1. Calcium Stones

Calcium stones more specifically calcium oxalate having a prevalence rate of 80% as compared to others. The main constituent of these stones being brusite, calcium stones formed other than calcium oxalate are calcium phosphate (15%-20%) and mixed (10-15%). Calcium oxalate (CaOx) crystals may be categorised into calcium oxalate monohydrate (COM, $CaC_2O_4.H_2O$) also known as whewellite and calcium oxalate dehydrate (COD, $CaC_2O_4.2H_2O$) or weddellite [4, 35]. Among the two, COM is considered thermodynamically more stable form as compared to COD. The COM form however, forms the more common cause of kidney stone formation since it has a greater affinity for renal tubular cells due to chemical nature of the formed crystals [35]. Factors contributing to oxalate stone formation include hypercalciuria (absorptive, resorptive, metabolic and renal leak), hyperuricosuria and hypercystineuria [4].

Other types of stones formed include struvite and magnesium ammonium phosphate, uric acid stone or uriate, cystine stones and drug induced stones [4, 35].

5.2. Struvite Stones

Struvite (magnesium ammonium phosphate) stones are the type of stones resulting due to UTI caused by urease- producing bacteria. The prevalence rate of the same being 10-15% among other forms of urinary calculi, these stones are also been referred to as triple phosphate or infection stones. Chronic UTI patients including those infected with *Protease mirabilis, klebsiella pneumonia, Pseudomonas aeruginosa and Enterobacter*, producing urease are more prone to such calculi. Urea cleavage into CO_2 and ammonia is mediated by urease thereby, increasing urine alkalinity (typically>7). This increasing alkalinity being unsuitable for phosphate as compared to acidic environment, leads to large stag horn stone formation resulting due to precipitation of this insoluble phosphate on insoluble ammonium products. This type of stone is more common among females [4, 24].

5.3. Uric Acid Stones or Urate

Accounts for about 3-10% of other renal calculi types. Purine rich diets especially animal protein diets including fish and meat results in decreased urine volume, hyperuricosuria and low urinary pH (pH<5.05) worsens calculi formation. People with gouty arthritis are also prone to such calculi. Uric acid stones being more common in males as compared to females, the most common cause for the same is idiopathic [4].

5.4. Cystine Stones

Comprising of less than 2% of all types of calculi, cystine stones seem to be larger in size, have a higher recurrence rate and are more likely to cause chronic kidney diseases as compared to CaOx stones. This is also categorised as a result of an autosomal recessive genetic disorder caused due to a mutation in one of the two genes in either SLC3A1 on chromosome number 2 (cystinuria type A) or SLC79A on chromosome number 9 (cystinuria type B) coding for dibasic amino acid transporter and major proximal renal tubule cystine. People with homozygous cystineuria are found to excrete more than 600 mill mole insoluble cystine per day. Other caused behind this type of stone formation being elevated urine L-cystine levels due to defect in reabsorption of filtered cystine. The condition is therefore exacerbated by its decreased L-cystine solubility thereby favouring facile crystal formation aggregating into stones often measuring in centimetres [4, 25].

The composition of stones include crystal as well as non-crystalline phases or the organic material (the matrix consisting of glycosylaminoglycans, lipids, proteins and carbohydrates) thereby promoting or inhibiting kidney stone formation. Main components of stone matrix include proteins (64%), nonamino sugars (9.6%), and hexosamine as glucosamine (5%), water (10%) and inorganic ash (10.4%) [4].

5.5. Randall's Plaque

Discovered in 1937 by a scientist named Randall, is the condition when the calcium oxalate stones are found attached to renal papillae containing interstitial apatite deposits. It was observed that the black calcium oxalate stones were found attached to white plaque composed of calcium carbonate/calcium phosphate as base. Derangement of papillary blood vessels, epithelial degeneration in collecting tubules, dense connective tissue and calcium deposition in the basement membrane are hypothesized to give rise to Randall's plaque. Also, it is found to contribute to calcium oxalate crystal lithigenesis; however, studies do not define this completely [28].

6. DEFENCE MECHANISMS AGAINST UROLITHIASIS

The pathogenesis of urolithiasis involves multiple mechanisms forming one of the most important reasons for restricted number of drugs available for its treatment. Oxygen-antioxidant balance in the kidney is observed to get perturbed by Reactive oxygen species (ROS) leading to cellular damage [9].

6.1. Reactive Oxygen Species (ROS)

ROS are produced by the renal epithelial cells when exposed to crystals such as CaOx, calcium phosphate and uric acid. ROS is also responsible for regulation of several macromolecular modulators of pathological calcification and plaque formation. It is also found responsible for oxidative stress development. The ROS productions is tightly controlled under normal conditions however, renal epithelial exposure to CaOx/calcium phosphate crystals and high oxalate leads to generation of excess of ROS that intern cause injury and inflammation. ROS and inflammation markers can be detected in urine of patients with stone and rats with artificially induced nephrolithiasis [9].

Conditions such as ROS production followed by renal cell injury and inflammation along with lipid peroxidation take place as a result of nephritic exposure to oxalate and calcium oxalate. The resulting loss in membrane integrity thereby facilitates calcium oxalate retention, promotes formation of collagen along with fibrosis and subsequent calculi formation. Reactive Oxygen Species formation result due to renal cell NADPH oxidase activation by the Renin Angiotensin System. Cytosolic phosholipase A_2 activation leads to arachidonic acid and lysophosphatidylcholine generation which increase ROS production leading to an increase in crystal formation and cell death. The obstruction caused due to stone formation results in renal colic and in such conditions, analgesic, antispasmodic activity and anti-inflammatory activity of smooth muscles are observed to play an important role in symptomatic relief from dysuria and renal colic. Antispasmodics are observed to play a significant role in easing out stone passage. Obstructed urine outflow resulting due to stone formation facilitates reduces the Glomerular Filtration Rate (GFR) leading to accumulation of nitrogenous waste (urea, uric acid and creatinine) in the blood. The Reactive Oxygen Species terminate phospholipase A_2 activation through NF- κB (nuclear transcription factor) [10].

6.2. Antioxidant Activity

Antioxidants leads to reduction of crystal and oxalate induced injury in animal models and tissue culture. Individuals with a kidney stone history were found to have significantly low serum levels of alpha-carotene, antioxidants, beta-cryptoxanthine and beta-carotene. An antioxidant rich diet is observed to reduce stone occurrence [9]. Oxidative Stress (OS) develops place due to overproduction of ROS and reduced cellular antioxidant capacities. It is due to down regulated antioxidant enzyme expression (catalase, superoxide dismutase, glucose-6-phosphate, glutathione peroxidise and phosphate dehydrogenase) along with radical scavengers (ascorbic acid, reduced glutathione and vitamin E) [11].

Studies also reveal that antioxidants from plant sources especially flavonoids and phenols are proven to ease stone formation in animal models as well as in humans. Plant polyphenols are an excellent group of antioxidants (exogenous) that act by hydrogen donating properties of their hydroxyl groups and electron donating properties to terminate free radical chain reactions. Other properties such as antimicrobial, anti-inflammatory as well as analgesic also play an important role against urolithiasis [9, 11].

7. INHIBITORS

Apart from this, Inhibitors of urinary calculi can be divided into Organic and Inorganic factors. Up-regulation of fibronectin, CD44, osteopontin, fetulin B and matrix-gla protein and down-regulation of inter-α inhibitor 1, 3 and 4, calgranulin B, prothrombin, and Tamm-horsfall proteins also stand responsible for urinary calculi [9]. However, Glycosylaminoglycans such as Hyaluronan are observed to promote nephrolithiasis. The reason behind such observation may be that HA being a major constituent of the extracellular matrix in the renal medullary interstitium and the pericellular matrix of mitogen/stress activated renal tubular cells. HA is said to be an excellent crystal binding molecule due to its size, negative iconic charge and ability to form hybrid gel like matrices. Therefore, HA bound crystal results in crystal retention in renal tubules (nephrocalcinosis) and in the calcified plague formation (Randall's Plaques) [12].

7.1. Osteopontin (OPN)

Osteopontin is one of the matrix proteins responsible for urinary calculi formation. OPN being multifunctional in nature is secreted by various types of body cells and is found responsible for a number of biological processes such as leukocyte recruitment, inflammation, cell injury and wound healing. The tissues possessing OPN include bone, kidneys, gall bladder, pancreas, lungs, breast, sweat and salivary glands. Strong OPN and mRNA expression is observed in distal renal tubular cells in kidney stone conditions and is observed to play a role in preventing crystal aggregation and it is also observed to play a role in adhesion of crystal to cultured epithelial cell [13].

7.2. Glycosylaminoglycans (GAG)

These are enzymatic products of proteoglycans and are observed to block the growth sites of calcium crystals thereby, inhibiting the growth and aggregation of these crystals. Also, they are found to prevent crystal adhesion with renal cells thereby obstructing renal stone formation [14]. Among the GAGs, urine contains sodium pentosan polysulfate (SPP), G871, G872, heparin sulphate, chondroitin sulphate and hyaluronic acid out of which only chondroitin sulphate is found absent in the stone matrix.

However, the concentration of GAGs in urine is insufficient to decrease calcium SS [15, 16].

7.3. Hyaluronan (HA)

HA, a linear GAG composed of multiple glucuronic acid and N-acetyl glucosamine (1, 4-GlcUA-1, 3-Glc-NAc-) $_n$, CD44 and CD168 acts as a receptor for HA. Glucosamine, a precursor for N-acetyl glucosamine forming one of the monosaccharide building blocks for HA. HA produced by HA synthase (HAS) proteins, located in the plasma membrane inner face, the site for its polymerization and stimulation, extruded into the extracellular space from the membrane. HA is divided into three mammalian genes- HAS1, HAS2 and HAS3 among which HAS1 is considered as the housekeeping HAS while HAS2 and HAS3 are regulatory enzymes. High molecular weight HA participates in a wide range of body functions including immune cell adhesion and receptor mediated signal transduction. The enzyme *Streptomyces* hyaluronidase (Hyal) is found to be HA specific digestive enzyme, others including HYAL1, 2 and 3coding for Hyal1, 2 and 3, CD44 receptor, β-glucuronidase and β-N-acetylglucosamineidase also perform HA catabolism. High activity of Hyal-1 and HA fragments are found in urine, HA playing a vital role in renal development. Studies show that the concentration of HAS2 mRNA and high molecular weight HA production up regulates in subconfluent culture while it down regulates in confluent ones. Hence, it was concluded that HA expression takes place in fetal human kidneys while it is absent in adult ones [12].

7.4. HA in Healthy Kidney

HA in healthy human kidney is undetectable in the cortex while is found in ample amounts in renal medullary interstitial. Subsequent breakdown of HA by Hyal in the medullary interstitium facilitates antidiuretic hormone mediated water movement out of the renal collecting duct. Generally, Hyal levels elevate while HA levels are reduced during low urine output while Hyal levels decrease and HA levels increase during water diuresis [12].

7.5. HA in Diseased Kidney

During inflammatory renal disease including interstitial nephritis, acute ischemic injury, autoimmune renal injury, acutely rejecting human kidney grafts, acute tubular necrosis and obstructed kidney and EG poisoning; HA levels are said to increase. During these diseased states, HA gets expressed in areas of the kidney such as corticointerstitium, and luminal surface of renal tubular cells; where it is absent in normal conditions. Sites of HA synthesis include interstitial fibroblasts, immune system cells – dendritic cells, lymphocytes and macrophages or renal tubular cells. Increased levels of HA are synthesized by the proximal tubular cells including HK-2 in response to scrape- surgery, high-D-glucose, IL-1β and bone morphogenic proteins-7 via intracellular signalling pathways such as mitogen activated protein kinase-dependent, NF-κB activated HAS2. HA is also observed to act as a binding molecule in for infiltrating monocytes [12].

7.6. Tamm Horsfall Proteins (THP)

In normal conditions, Tamm-Horsfall protein is considered as the most abundant glycoprotein in the mammalian urine after albumin under normal conditions (Aggarwal et al., 2013). Also known as uromucoid, this glycoprotein is synthesized in the thick ascending limb of loop of Henle. THP is found to be involved in the pathogenesis of cast nephropathy, tubulointerstitial nephritis and urolithiasis. It is also observed to inhibit viral hemagglutination. THP is also observed to act as the first line of host defence in against bacterial infection and renal stone formation. THP appears to aggregate more easily in diabetic patient's possible reason behind this may be altered glycosylation [15].

7.7. Urinary Prothrombin Factor

Discovered from freshly precipitated CaOx crystals and taking up stain of glycoprotein, it was incorporated selectively into the crystals with excessive quantities as compared to others. It is found to be the principle constituent of CaOx stone due to direct inclusion into crystalline architecture as compared to secondary product of tissue injury. This is hence found to be the most prominent inhibitor of CaOx stones. It is also known as potent inhibitor of CaOx crystal aggregation in undiluted urine [17].

7.8. Fibronectin

Fibronectin is a α_2-glycoprotein, multifunctional in nature, spreads throughout the extracellular matrix and body fluids. The major sources of FN include hepatocytes, fibroblasts, vascular endothelium and glomerular cells. However, the three types of FN producing cells in the kidney include-mesangial, epithelial and endothelial that intern makeup the glomeruli and tubular cells. Over secretion of FN takes place in the renal tubular cells due to stimulation of CaOx crystals and plays a role in inhibiting the aggregation of CaOx crystals thereby preventing there adhesion to renal tubular cells [18]. Characteristics of FN due to its adhesive action include morphogenesis, metastasis and wound healing. The property of its wound healing proven helpful in repairing the renal tissues damaged due to exposure to CaOx crystals [19].

7.9. Inter-α-Inhibitor (IαI)

Inter-α-inhibitor is a heterotrimer glycoprotein belonging to a group of kunin possessing kinitz-typeprotein superfamily having a common structural element and serine protease inhibiting ability. Synthesized in liver, this inhibitor is composed of a combination of H1 (60kDa), H2 (70kDa) and H3 (90kDa) heavy chains covalently linked to a light chain known as Bakunin via chondroitin sulphate bridge. Bikunin expression is mainly seen in proximal tubular cells and thin descending segments near the loop of Henle. Increased expression of this bikunin mRNA is observed in renal epithelial cells on exposure to oxalate and CaOx crystals. As a result of this bikunin is observed to have inhibitory activity against CaOx. This activity of inhibition takes place at the carboxyl terminal of the bikunin fragment of the IαI. It is observed to inhibit nucleation and growth of CaOx crystals. This inhibitory activity was however destroyed by pyranose treatment thereby, indicating the involvement of peptide chain rather than the chondroitin one. Apart from this, it is also observed to play a role in adult respiratory distress syndrome, septic shock treatment [20].

7.10. Renal Lithostathin

A glycoprotein secreted in pancreatic juice and synthesized by the acinar cells. Pancreatic juice found to be supersaturated with calcium and bicarbonate ions, lithistatin however, is observed to play an important role in inhibition of growth of calcium carbonate crystal. A protein actually present in urine of healthy individuals and is immunologically related to lithostathine and in renal stones renal lithostathin. It is named so due to its structural and functional similarity pancreatic lithostrathine. Renal lithostathine contributes in controlling calcium carbonate crystal growth. Reports also suggest the presence of calcium carbonate crystals in renal stones might also be present in early steps of stone formation. Such crystals seem to provide appropriate substrate required for heterogenous nucleation thereby, promoting CaOx crystallization [20].

7.11. CD44

A glycoprotein, multifunctional in nature and present in the extracellular matrix is observed to be involved in migration, adhesion and cell-cell adhesion. Expressed in a number of different types of mammalian cells, it is also found in association with crystal adhesion with renal epithelial cells. It is also found to play a major role in CaOx crystal binding during the process of wound healing. CD44 and HA upregulates during inflammation and injury. At confluence, HA being undetectable, CD44 is expressed at the basolateral membrane. Proliferating cells being receptive to CaOx adhesion lose their property when cells become confluent. Therefore, it was concluded that crystals are not bound to intact epithelium due to absence of a pericellular matrix and attachment of cell depends on CD44, OPN and HA expression [20].

7.12. Matrix Gla Protein (MGP)

A natural vascular calcification inhibitor, it is a vitamin K dependent extracellular matrix protein, isolated initially from bone, is also expressed in heart, lung, vascular smooth muscle cells of the blood vessel wall, and kidney. MGP is observed to be involved in differentiation, growth, and growth of apoptosis and increase cell density in normal kidney cells. MGP being molecular determinant regulates vascular calcification of the extracellular matrix. MGP in the apical membrane of tubular epithelial cells binds directly to crystals found in ascending thick limbs of loop of Henle, and DCT in hyperoxaluric rats. Exposure of renal tubular cells to calcium oxalate was followed by up regulation of MGP mRNA expression. Therefore, MGP was observed to play a cytoprotective role to maintain cells' survival and inhibit crystal retention under crystal and oxalate exposure [20].

7.13. Nephrocalcin (NC)

Major components of nephrocalcin include acidic amino acid residues (glutamic acid and aspartic acid contribute to about 25% of these acidic amino acids). The γ- carboxyglutamic acid (Gla) residues are found to contribute in restriction mechanism of calcium oxalate crystallization. The site of expression of NC involves proximal tubules and the thick ascending limb of loop of Henle (Coe et al., 1993).

Apart from the inorganic inhibitors, some organic inhibitors also play a role in inhibiting urolithiasis. These include:

7.13.1. Citrate

A tricarboxylic acid, circulating in blood complexed with calcium, sodium and magnesium at7.4 pH derived from endogenous oxidative metabolism. Citrate is known to perform stone inhibiting action in urine and is found effective against calcium oxalate and phosphate stones. It is found to alter both COM and calcium phosphate crystals. Its complex with calcium leads to reduction of total calcium content in urine leading to reduced frequency of calcium stone formation. It is also found to increase inhibitory activity of other inhibitors including THP and reduce OPN concentration (important part of matrix). Additionally, it is found to increase urinary pH leading to alkalisation, formation of calcium-citrate-phosphate complex formation, uric-acid solubility thereby preventing calcium oxalate salting out process [21].

7.13.2. Magnesium

Being the forth most abundant mineral in the human body, magnesium is majorly found in bones. Dietary magnesium is excreted through the kidney while it is absorbed through the small intestine only 1% of which is being circulated in the blood. 2mmol/L magnesium was found to reduce 50% of particle number in a super saturated solution of CaOx the reason behind this is that magnesium forms complexes with oxalate thereby reducing supersaturation. Oral magnesium intake will lead to oxalate binding leading to reduced absorption and excretion of urinary oxygen.

This procedure may be analogous to oxalate calcium binding taking place in the gut. Also, magnesium deficiency patients supplemented with oral magnesium leads to increase in excretion of citrate in urine [21].

7.13.3. Pyrophosphates

The normal range of pyrophosphate content in the blood (20-40μM) estimated to be abrupt to inhibit CaOx and CaP crystallization. Pyrophosphates along with diphospahtes are observed to inhibit CaP precipitation while diphosphates are also observed to inhibit apatite crystal growth. Pyrophosphates are also found effective in reducing calcium absorption in the intestine the action being mediated by 1.25 (OH)$_2$ – vitamin D formation. Oral orthophosphate administration is shown to be less affective against stone recurrence [21].

7.13.4. Phytate

An inverse relationship between dietary intake of phytate (IP6) and calcium urolithiasis is observed. Additionally reports also reveal an inverse relationship between urinary excretion of this myoinositol hexaphosphate and stone formation. The mechanism behind this property has been endorsed to its kinetic role as a calcium oxalate (calcium and phosphate) crystallization inhibitor. Phytate though not observed to perform thermodynamic inhibitory effects on crystallization of calcium stone forming salts in urine. Using comprehensive animal models it has been demonstrated that phytate inhibits crystal aggregation kinetics. Results of theoretical remodelling lead to identification of an unbound species HPhy^{-11} playing a role in the latter mechanism [22].

7.13.5. Genetic Factors

Genetic factors such as Calcitonin receptor (CALCR) rs1801197 and CALCR rs1042138 are reported to be associated with pediatric and adult urolithiasis and also with kidney stone reoccurrence. Apart from this, mutations in AGXT, GRHPR, HOGA1 three types of primary hyperoxalirias are reported to cause overproduction of exdogenous oxalate leading to kidney stone disease due to genetic defect in glyoxylate metabolism. Renal damage and calcium oxalate deposition were observed as a result of hyperoxaluria.

Gene candidates postulated by genome analysis include Adenylyl cyclise, Claudin 14 and calcium sensing receptors are thought to be associated with hypercalciuria and urolithiasis. However, phenotypic evidence of the same is not available.

Also, studies on single nucleotide polymorphism (SNP) with genetic disorders explore gene such as claudin family(CLDN14,16,19), SLC family (SLC34A1), Ca^+ transporter (TRPV5), Isozyme of alkaline phosphatise (ALPL), Calcium sensing receptors(CASR), and some genes from glyoxalate cycle (ATG and HPR).

Conditions such hyperuricosuria, hyperuricemia, gout and kidney stone formation are resulted due to monogenic mutation in a locus on chromosome 10q21-22 causing UA- nephrolithiasis, further found as gene encoding zinc finger protein.

An inherited defect in dibasic amino acid exchangers SLC3A1 and SLC7A9 caring out reabsorption of cystine in urine are found to be associated with cystine stone formation. These genes are said to promote cystine reabsorption along with reabsorption of other amino acids such as ornithine, arginine and lysine. A defect in which may lead to cystine stone formation [20].

8. PROMOTERS

Components contributing to calculi formation along with its growth and aggregation are referred to as promoters. These promoters include low urinary volume and pH, components such as calcium, oxalate, sodium and urates are the major promoters of lithogenesis and are found to be well established for their capacity of inducing supersaturation, crystal nucleation and aggregation (Goldfarb, 2011).

8.1. pH

The normal range of urinary pH is around 7. An alteration in the same observed usually during acidosis, metabolic syndromes (Kohjimoto et al., 2013). These alterations change the physicochemical nature of urine thereby restraining the actions of inhibitors and favouring stone formation (Lieske et al., 1994).

8.2. Oxalate

Oxalate forms one of the primary risk factors contributing to calculi formation. Increased dietary or idiopathic hyperoxaluria induces supersaturation leading to crystallization in addition to calcium. This process is followed by spontaneous nucleation of CaOx crystal along with aggregation and retention (Gupta et al., 2011).

8.3. Calcium

Hypercalciuria decreases the inhibitory activity of a number of urinary components thereby favouring formations of crystals. Apart from them, ionic components such as sodium and phosphate are also found to contribute towards crystallization (Gupta et al., 2011).

9. Mechanism of Kidney Stone Formation

The process of crystallization is regulated strongly by inhibitor and promoters of kidney or associated polymers. Mechanism of stone formation involves 4 basic processes:

- **Nucleation:** Considered as a crucial step in lithogenesis; i.e., a solution found with solid crystals. The process of nucleation initiates due to supersaturation; a state of liquid consisting of solutes whose concentration exceeds the dissolvable limit under normal circumstances. When this saturation point is reached, the constituents of the urine begin to precipitate this is achieved when no more solute can be added to the solution and hence, the process of crystallization begins denoted as thermodynamic stability product (ksp). However, metastable state is defined as the state in which large amounts of solute above the ksp can be dissolved by the urinary inhibitors of crystallisation. The product formation occurs at a point known as KF also known as the threshold point where no more solute can be dissolved in the solution. In case of pure solution, homogenous nucleation occurs while secondary nucleation occurs due to occurrence of crystal nucleation on a pre-existing crystal surface. The other type of nucleation, known as heterogeneous nucleation or epitaxy, is obtained when one type of crystals promote formation of the second one an example being the process of formation calcium oxalate crystals promoted by uric acid crystals. Therefore, urine being solution mixture of a variety of components, it is considered as a heterogeneous mixture [21].
- **Growth:** After the crystal nucleus has achieved relative supersaturation and critical size, addition of new crystal components to the nucleus leads to overall decrease in free energy. This process called crystal growth, is one of the prerequisites for particle formation leading to stone formation; also crystal growth and aggregation play an important role in each step of stone formation. The crystal growth rate being low, the tubular fluid transit time through the kidney amount to several minutes. A single particle achieving a pathophysiologically relevant size by the process of crystal growth alone seems to have relatively less probability. This criterion doesn't seem to change much even if the growth rate proceeds at an uninhibited rate of 2mm per minute [21].

- **Aggregation:** Crystallization being a common process occurring even in healthy individuals may be regarded as harmless when individual or small aggregates of calcium oxalate are excreted while the alarming stage is when large aggregates of such stone forming salts are excreted. Crystal aggregation forms a crucial step in stone formation and is determined by the balance of aggregating and disaggregating forces [22].
- **Retention:** Crystal retention results due to interaction between these crystals and renal epithelium specifically in the renal tubular region. The composition of glomerular filtrate (after proximal reabsorption), and composition of the epithelial surface decides the process to be employed for crystal retention. However, parts such as distal tubules, ureters, bladder collecting ducts, possess a natural defence mechanism against crystal retention that is against crystal retention, this when compromised, leads to significant crystal deposition and stone formation [23].

10. Treatment Methods Employed for Urolithiasis

10.1. Diagnosis

The primary diagnostic methods employed for the treatment of urinary calculi include clinical, physical and radiological evaluation.

10.2. Clinical Examination

The clinical examination comprises of symptoms responsible for urolithiasis. These symptoms differ according to the stone size. These include dysuria, hematuria, stanguria and pollakiuria. The symptoms for small uroliths include abdominal pain, post renal azotemia, with anorexia, nausea, bladder distention along with distention (Langston et al., 2008).

10.3. Physical Examination

The physical examination include of urolithiasis include palpation of the abdomen to rule out the stone and thickened bladder walls in case of cystic calculi. Stones in ureter and urethra can be analysed by rectal examination (Langston et al., 2008).

Hypercalcimia and metabolic acidosis are the haematological and biochemical variations associated with urolithiasis. Apart from this, infectious stones may show symptoms such as increased leukocyte count along with abnormal concentration of urine constituents. pH is yet another indicator as acidic pH is observed in the case of urate stones while the pH turns alkaline during struvite stones. Analysis of the type of stone is another parameter to analyse the recurrence of rare stones including cysteine and struvite stones (Langston et al., 2008).

10.4. Imaging (Radiological Examination)

10.4.1. Unenhanced Helical Computed Tomography

A highly sensitive (>95%) and specific technique (>96%), introduced in 1990s, is considered as the first choice for initial and subsequent patient evaluation in case of urolithiasis. Being non-contrast in nature, it offers a number of advantages including high sensitivity and specificity for stone detection, easily available, faster acquisition speed, and intravenous contrast administration is not required [29].

10.4.2. Multi Detector CT Scan

Introduced in 1998, MDCT, is found to be highly sensitive (up to 98%) and specific (96-100%) in diagnosis and is considered as imaging technique of choice for initial evaluation in patients suspected with kidney stone. The advantages include ability to detect extra urinary pathologies including appendicitis, diverticulitis or gynaecological pathologies such as haemorrhagic cyst or ovarian torsion mimicking renal colic. Such advantages play a crucial role in managing patients with acute abdominal pain imitating renal colic in emergency departments [29].

10.4.3. Dual-Energy CT Scans

This technique is said to show accurate determination of stone composition. This involves simultaneous scanning two different energies leading to tissue characterisation [29].

10.4.4. Excretory Urography

Used for analysing the calculi present in the upper urinary tract. The images obtained reveal a clear ureter obstruction in case of abrupt termination of contrast along pelvic enlargement and reduction in distal enhancement (Botsikas et al., 2014).

10.4.5. Antegrade Pyelography

This technique is an ultrasonographic or fluoroscopic imaging technique used to obtain better images of stones present in the ureters and collecting ducts. This is achieved by injecting dilated renal pelvis with contrast material (Naitoh et al., 2015).

10.4.6. Abdominal Ultrasonography

This technique is employed to visualize both radiopaque and non-radiopaque calculi. However, its sensitivity towards ureter stone is about 77%; it can be combined with other techniques such as survey radiography to improve the sensitivity to 90% (Nicolau et al., 2015).

The above techniques are considered important for diagnosis, planning treatment and treatment monitoring. Imaging techniques are also employed to deduce the size and location of the urinary calculi.

The methods of treatment involve surgical management and drug administration.

10.5. Surgical Management

The surgical management include surgical removal of urinary calculi having diameter more than 7mm. Major factors kept in mind while deciding the procedure of treatment include methods that reduce the symptoms, complete stone removal and prevention of stone recurrence (Bartoletti and Cai, 2008). The techniques hence employed include:

10.5.1. Percutaneous Nephrolithotomy
This surgical procedure depends on the stone size. The procedure however, involves establishment of a percutaneous working tract for fragmenting and removal of stones using ultrasound, laser, or pneumatic lithotripsy. Such techniques are usually employed to analyse large sized stones in the upper urinary tract of children [1].

10.5.2. Extracorporal Shockwave Lithotripsy
One of the widely practiced techniques for the treatment of urolithiasis. Employed for the treatment of stones located in 3 areas namely caliceal (42.6%), renal pelvis (27%), and ureteral (30.4%). The recurrence risk after such treatment was reported as 27% including anabolic and metabolic abnormilities. The stone free rate after the first treatment was observed to be 60% which increased to 68% after second treatment. Larger stones covering an area of 40×30 mm, are treated employed using a combination of percutanious lithotripsy and extracorporal shock wave lithotripsy [3, (Kupajski et al., 2012; Grasso and Goldfarb, 2014)].

10.5.3. Ureteroscopy
A widely employed method for the treatment of small stones due to lower costs and recurrence rates. In this treatment a small scope (similar to a flexible telescope) is inserted into the ureter and urinary bladder. It is also employed for the non-surgical removal of urethral stones. This procedure occasionally employs a stent insertion into the ureter thereby, helping it to remain open. However, fragmentation of the stones using helium laser device ureteroscopy is more assured [2].

10.5.4. Open Surgery

This technique is employed for the treatment of large stones in children often requiring multiple endoscopic procedures. Children with congenital abnormalities of urinary system where especially the cases in which operative positioning is not advised. Have high clearance rate [4].

10.5.5. Ureterorenoscopy (URS)

Involves insertion of an ureterorenoscope through the bladder, only drawback being the size of ureteric orifice may be insufficient for the treatment in smaller children [4, Jung and Osther, 2015].

10.5.6. Different Contrast Lithotripsy

This technique is employed for treatment of stones of size less that 4mm that are usually not detectable through ESWL especially in children. A combination of ultrasonography and pneumatic lithotripsy are often used along with ureterorenoscopy for better results [4].

10.5.7. Laparascopic and Robotic Surgery

These techniques are gaining popularity for the treatment of urological condition requiring a surgical approach but are less common for paediatric stone treatment [4].

10.5.8. Medical Expulsion Therapy

Conservative medication based method for stone management. Involves conservation of medications to facilitate and accelerate the spontaneous passage of ureteric stones. Corticosteroid hormones, non-steroidal, inflammatory agents, calcium channel blockers and alpha-adnergic blockers have been used in the conservation management of stone disease [4].

Mechanism of action employed by the drugs for the treatment of urolithiasis.

Pathophysiologic and Therapeutic Aspects of Renal Calculi

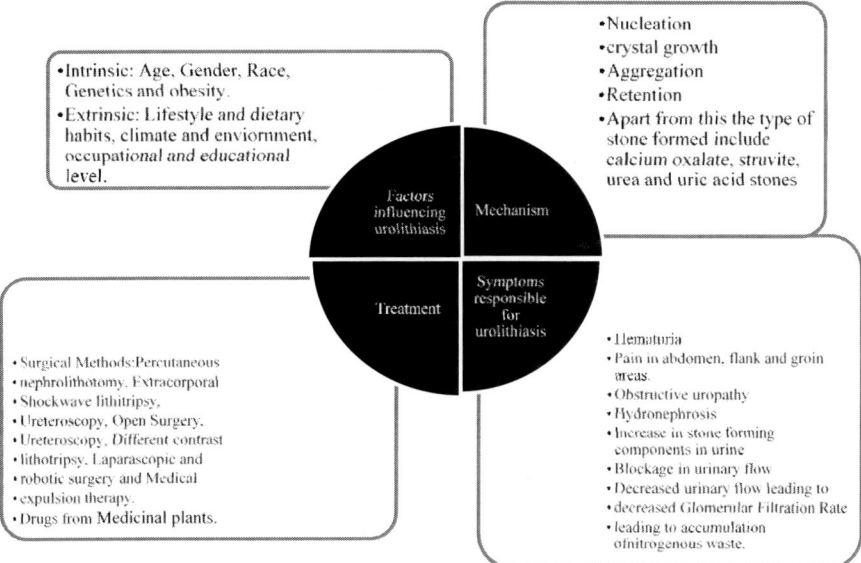

Figure 2. Urolithiasis: An overview of the factors, mechanisms, symptoms and treatments of urolithiasis.

Drugs exhibiting multiple defensive mechanisms act as a breakthrough for minimizing tissue injury due to various human diseases. Multiple phytoconstituents present in herbal medicines exert their beneficial effects via multiple mechanisms including:

- These drugs increase certain parameters including urine volume, pH and diuretic (anticalcifying) activity thereby easing out the spontaneous process.
- Helps relieve the binding mucin of calculi (lithotropic activity)
- Affects the crystal nucleation, aggregation, and growth (inhibition of crystallization) by balancing the inhibition and promotion of crystallization process in urine.
- Leads to improvement in renal function.
- Regulate oxalate metabolism.
- Prevent reoccurrence (Antioxidant activity); improve renal tissue antioxidant status and integrity of the cell membrane.

- Prevents renal calculi reoccurrence by regulating crystalloid colloid imbalance and improving renal function.
- Exerts antimicrobial activity by exerting significant anti-infective action against major causative agents.
- Inhibition of ACE and Phospholipase A2.
- Show remarkable improvements in urinary calculi symptoms including burning micturition, pain, and hematuria (Anti – inflammatory and analgesic activity) [11].

11. THERAPEUTIC EFFICACY OF SOME MEDICINAL PLANTS FOR THE TREATMENT OF UROLITHIASIS

11.1. *Abelmoschus Moschatus*

A.moschatus is an herbaceous plant belonging to the family Malvaceae; commonly found in India, china, tropical Asia and pacific islands. Antiurolitiatic activity of A.moschatus was examined in zinc rats by incubating them with the seed extracts of the plant. For this analysis, urolithiasis in the rats was introduced into the urinary bladder through surgical implantation of zinc discs. Postsurgical recovery was followed by administration of different doses of methanolic (MAM) and chloroform (CAM) of A.moschatus seeds were given to disc implanted rats orally for a period of 7 days. The antiurolithiatic activity was analysed by serum and urine analysis of these rats performed along with the evaluation of various stone dimensions. After implantation, the urinary output of the rats decreased significantly while after treatment with the extracts, the normal frequency of urine was retained. Also, extract supplementation caused significant GFR (Glomerular filtration Rate) and total urinary protein extraction improvement. Other features of these extracts include prevention of increased serum creatinine, uric acid and blood urea nitrogen levels also reducing calculi deposition surrounding the implanted disc. Significant antiurolithiatic potential was observed in higher doses only of chloroform (CAM) while it was observed in all the three doses of methanol (MAM). Therefore, the seeds were shown to exhibit antiurolithiatic activity the underlying mechanism for the same being mediated collectively through antioxidant, diuretic and free radical scavenging effects of the plant [30].

11.2. *Acorus Calamus*

Acorus calamus (Family: Araceae) (EEAC) commonly called "sweet flag" is a well known Indian medicinal plant commonly found in the marshes (wild or cultivated) ascending from 2200m in Himalayas, possessing a variety of ailments.[31, 32] The medicinal properties of A.calamus include treatment of insomnia, remitten fever, hysteria, memory loss, necrosis and epilepsy. The rhizome of the plants is reported to possess diuretic, antioxidant properties and nephroprotective effect. Oral dose of ethylene glycol administrated male wistar rats were analysed for the antiurolithiatic effect of EEAC using Cystone as the reference drug. The antiurolithiatic effect of EECA was assessed by analysing rat biological samples (urine, serum and kidney homogenate) for levels of various urolithiatic promoters and renal function after a 28 day treatment. When compared to the standards such as furosmide administration, EEAC administration resulted in significant increase in urine volume and urinary excretion of Na^+ and K^+ electrolytes. Also, a decrease in excretion and deposition of various urolithiatic promoters as compared to standard results observed in case of cystone administration were observed in ethylene glycol induced model. Apart from preventing renal function impairment, EEAC's antiurolithiatic mechanism is also mediated through diuretic and nephroprotective actions of active rhizome compounds [31].

11.3. *Aerva Lanata*

Pashanabheda plants are antiurolithiatic drugs traditionally used in the Indian medicinal system. *A.lanata* is a Pashanabheda plant belonging to *Amaranthane* family [33]. In traditional medicine, *A.lanata* is commonly used in cough, stangury (painful and slow urine discharge), urolithiasis and headache [34]. Commonly found in Western Ghats of Khanapur region of Belagavi district of Karnataka, this plant also seems to possess significant antiurolithiatic activity. Rats treated with ethyl acetate and n-butanol extracts of quecertin and betulin showed significant increase in urine volume. Significantly enhanced calcium, phosphate and oxalate levels, increased magnesium levels and reduced calculi size were observed in urine microscopy. SEM (scanning electron microscopy) of kidneys of treated animals revealed significant reduction of renal calculi along with reduced BUN and creatinine levels. *A.lanata* therefore, showed mild antiurolithiatic and diuretic activities [33].

11.4. *Betula Alba* (Betulin)

B.alba from the family *Betulaceae* apart from treating urolithiasis found to be helpful in treating arthritis, gout, boils, fever, headache, worms and rheumatism. It is observed to possess anti-inflammatory activity. The leaves of *B.alba* being diaphoretic and antibacterial in nature are employed for treating gonorrhoea, diarrhea, dysentery, cholera, etc. However, their essential oils are used to in case of eczema, psoriasis and hair loss. The study on betulin (the main phytochemical of B.alba) revealed its high antioxidant and inhibitory effects against CaOx crystals. The results also demonstrate significant scavenging activity against DPPH, NO, and superoxide radicals when compared to standards such as L-ascorbate (L-AA). In silica studies reveal that Betulin and L-AA show strong affinity towards matrix metalloproteins [35].

11.5. Biophytum Sensitivum

B.sensitivum plant belongs to *oxalidaceae* family and is distributed in the tropical parts of Asia, America, Africa and Philippines. Commonly found in wetlands, at low and medium magnitude and under tree shades, this plant is widely distributed among hotter parts of Nepal, Malaysia, India, Indonesia Thailand and Sri Lanka [36, 37]. The antiurolithiatic effect of methanolic whole plant extract of *B.sensitivum* is reported against zinc disc implantation induced lithiasis and ethylene glycol induced and ammonium chloride induced lithiasis in rats. Antiurolithiatic activity of the prepared extract was assessed by analysing biochemical changes in serum, urine and kidney homogenate along with analysis of histological changes in kidney tissue. Ethyl acetate fraction supplementation prevented biochemical alterations in urine, elevated serum uric acid, creatinine and blood urea nitrogen levels. Reduced levels of calcium, phosphate and oxalate in lithiatic rat's kidney homogenate of were also an outcome of this treatment. The ethyl acetate fraction was also observed to show significant decrease in accumulation of calcium oxalate deposits, histological changes in kidney tissue and peroxidation activity. The plant is therefore, observed to show diuretic, antioxidant, and nephroprotectice effects on decreasing concentration of urinary stone forming constituents [37].

11.6. *Carica Papaya*

C. papaya, belonging to the family *Caricaceae*, is commonly available as Papaya fruit; native to tropical regions of the world [38, 39]. This fruit along with other parts of the plant is well known for its medicinal properties under the traditional system of medicine. The traditional uses of the extracts of leaves, roots, fruits and seeds of *C.papaya* include its use as an antioxidant, anti-inflammatory, diuretic, antihypertensive, wound healing, anti ulcer, hypoglycaemic and hypolipidmic. Antiurolithiatic activity of aqueous and alcoholic extracts of *C.papaya* was examined on ethylene glycol induced urolithiatic rats. EG administration resulted in hyperoxalurea and increased renal excretion of calcium and phosphate. A significant reduction in elevated urinary oxalate, showing a regulatory action on endogenous oxalate synthesis was observed on treatment with aqueous and alcoholic extracts of C.papaya. A significant reduction in of stone forming constituent levels was also observed as a result of curative and preventive properties of alcoholic and aqueous extracts of *C.papaya* [38].

11.7. *Cissampelos Pareira*

C.pareira belonging to the family *Menispermacia* is a commonly found shrub in India. Due to its antiseptic properties of *C.pareira* is used for treating urinary tract infection. Other uses of this plant comprise of its use for the treatment of muscle inflammation, rheumatism, snakebite, diarrhea, menstrual problems and dysentery. This herbal medicine is also found effective as a tonic, diuretic as well in reducing pain and relieving fever. The treatment with alcoholic extract of *C.pareira* given at three different doses significantly reduced urinary calcium, uric acid level. An increase in magnesium and decrease in serum creatinne and creatinine levels was observed. Less tissue damage in case of rats treated with higher doses was also observed [40].

11.8. *Costus Spiralis*

C.spiralis commonly known as cana-de-macaco belonging to the family *Zingiberaceae;* is widely distributed through regions of Brazil and Sao Paulo [41, 42]. Medicinal uses of this plant include there use as diuretic, and treatment of urinary tract infections. Reduced calculi growth on administration of the extract 4 weeks after surgery was observed. Hypertrophy of the smooth musculature was however, not prevented. No difference among the experimental groups among the contractile responses of isolated urinary bladder preperations to the muscarinic agonist bethanecol in the absence or presence of extract or atropine [41].

11.9. *Crataeva Magna* Lour. Bark

Crataeva magna Lour. Bark, commonly known as Baruna and belonging to the family Capparaceae, is widely distributed along tracks of Himalayas extending towards Bihar, Assam hills, North Bengal, Meghalaya, Arunachal Pradesh and other parts of India [43, 44]. *Crataeva magna* Lour. Bark is traditionally used as antispasmodic, antihypertensive, anti-inflammatory, treatment of kidney stones and as analgesic. The extract was observed to significantly reduce the serum calcium levels in lactose + ethylene glycol induced urolithiatic model. While, the ethylene extract significantly reduced uric acid, serum creatinine and calcium, urine oxalate and kidney weight levels in both the models compared to toxic group [43].

11.10. *Gokshuradi Yog*

Gokshudi Polyherbal Formulation (GPF) is proven to explode kidney stones in naturally occurring urolithiasis. GPF is a combination of five different medicinal plants including dry fruits of *Tribulus terrestris* L., seeds of *Hygrophyla spinisa T. Anders* and roots of *Solanum anguivi Lamk, Ricinnus communis L and Solanum surattense Burm*. The fruits of *T.terrestris L* are used as anti bacterial, anti hypertensive in coronary heart diseases; stimulate melanocyte proliferation, treating urinary diseases. *H.spinosa I*T.Anders has been used against jaundice, rheumatism, hepatic obstruction, diuretic, pain, inflammation, aphrodisiac and treating urinary disorder, possessing hypoglycaemic activity, diuretic, hepatoprotective. *R.communis* possesses anticancer, antiprotozoal and antidiabetic activities. *Solanum anguivi* (Lam.) also called *Solanum indicum* sp. distichum usually used for culinary purposes, in many parts of Africa, India and Arab peninsula, it is a reservoir of starch, calcium, vitamin A, phosphate and ascorbic acid. *Solanum xanthocarpum* belonging to the Family Solanaceae is popularly known as Indian night shade found throughout India, mostly as a weed in dry places, roadsides and wastelands. It isusually employed as an antipyretic and expectorant, other medicinal properties include treatment for asthma, catarrahal fever and chronic cough. Its root decoction is used as an effective diuretic and febrifuge. Significant antioxidant activity against 1, 1-diphenyl-2-picrylhydrazyl free radical inhibiting lipid peroxidation exhibited by Gokshuradi Polyherbal aqueous extracts (GPAE) was observed in *in-vitro* experiments. Higher doses of GPAE lead to diuresis along with saluretic effect thereby revealing decrease in oxalate synthesizing and increased antioxidant enzyme activities [45].

11.11. *Holarrhena Antidysentrica*

H.antidysentrica (Ha.Cr) belongs to the family Acanthaceae commonly found in moist areas, swampy grounds and along the banks of stagnant and fresh water ditches mixed with sledges and marshy grasses of tropical Africa and Southern continental Asia including India and Myanmar along with Sri Lanka [46, 66, 67]. Commonly known as Kutaza, Kurchi or Bitter oleander; Ha.Cr has a traditional use in urolithiasis treatment, other than this its medicinal uses include treatment for jaundice, diarrhea, boils and bronchitis, leprosy, body pain, cough, skin diseases, gall stones, arthritis and fistula [46, 47, 48]. Its crude aqueous-methanolic extract indicated a concentration –dependent inhibitory effect on the slope of aggregation. Crystal size reduction and transformation from Calcium oxalate monohydrate (COM) to calcium oxalate dehydrate (COD) was also observed in metastable solution. Concentration dependent antioxidant effect against 2, 2-diphenyl-1-picryl-hydrazyl (DPPH) free radicals and lipid peroxidation in rat kidney tissue was also observed. Ha.Cr showed reduction in cell cytotoxicity and LDH renal epithelial cells (MDCK) on oxalate exposure and COM crystals [46].

11.12. *Hygrophila auriculata (Schum.) Hiene*

H.auriculata belongs to the family Apocynaceae usually found in mountain areas of India and neighbouring countries [49, 50]. The medicinal properties of H.auriculata include its use for treatment of fever, diabetes mellitus, piles, asthma, dysentery, diarrhea, malaria, eczema and bronchopneumonia [50]. Extracts and bioactive compounds of H.auriculata have been found to possess antimicrobial, nephroprotective, hepatoprotective, antitermite, anticataract, antioxidant, hematopoietic, etc. Treatment of ethylene glycol induced mice with alcoholic (n-butanol) extract of *H.auriculata* significantly reduced urinary calcium, oxalate and serum uric acid levels with increase in urinary magnesium. It also reduced serum oxalate and calcium in the kidney thereby, reducing deposition of stone forming compounds [49].

11.13. *Lantana Camara*

L.camara, popularly known as Spanish flag or West Bengal Lanata and Raimuniya (common Hindi name) belongs to Verbenaceae family [51]. The Ayurvedic names of this plant include Chaturaanga and Vanachchhedi; L.camara is abundantly found in tropical and sub-tropical areas and also used for ornamental purposes other than medicinal ones in India. The medicinal implications of L.camara include its use against chicken pox, high blood pressure, cancer, measles, eczema, fevers, carminative, diaphoretic, bilious fever, tetanus, tumors, ulcers, swelling etc. [51, 52]. Ethanolic extracts of roots (ELC) and oleanolic acid (OA) extracts of L.camara were used against zinc disc induced urolithiasis in rats were examined. Treatment with ELC and OA lead to significant reduction of calcium output along with reduced deposition of crystals on implanted zinc discs and non-significantly increased urine output. Increase in urine output took place as a result of activation of mascarinic receptors present in the bladder along with other mechanisms [51].

11.14. *Lepidagathis Prostrate*

L.prostrate from the family Acanthaceae is a Pashanbhed plant that grows abundantly in the Western Ghats region of India. It is a commercially available Ayurvedic drug widely employed as lithotripic and diuretic. Other than this, it is traditionally used for the treatment of dysuria, urinary calculi, fever, polyurea, and cardiovascular diseases. Plant extracts in Methanol (LPM), Petroleum ether (LPPE), Ethyl acetate (LPEA), n-butanol (LPBU), and aqueous (LPAQ) were prepared. Out of which LPEA was found to exhibit the highest dose-dependent CaOx nucleation and aggregation inhibition that was found to be significantly better than Cystone. Phenol and flavonoid rich polar LPBU fraction was correlates with its highest antioxidant potential in DPPH, ABTS, nitric oxide scavenging and iron chelating activities [9].

11.15. *Paronychia Argentea*

P.argentea belongs to the family Illecebraceae and is popularly known as Arabian tea is covers large parts of East Algeria and Mediterranean area [53, 54, 55]. The aerial parts of the plant prove to be of medicinal use under the Algerian system of medicine as diuretic and treating renal diseases especially renal diseases, antimicrobial activity and hypoglycaemic activity. P.argentea is used as an analgesic, in anorexia, stomach ulcers and flatulence. The plant extracts proved to be advantageous in preventing retention of urinary stone by reducing renal necrosis thereby inhibiting crystal retention. Elevated serum creatinine and blood urea levels (nephroprotective effect) were observed on treatment with butanolic extract (BPA). However, no significant alterations in biochemical, haematological and morphological parameters were induced due to P.argentea [53].

11.16. *Pergularia Daemia*

P.daemia belongs to the family Asclepiadaceae also known as Uttaravaruni in Sanskrit is a perennial herb found abundantly on Indian roadsides. The traditional uses of the plant include its use as antipyretic, laxative, anthelmintic, expectorant and in infantile diarrhea. Apart from this, the plant has also been used as anti-inflammatory, analgesic, anti-pyretic, anti-diabetic, antifertility, antibacterial, wound healing and hepatoprotective activity. Urinary excretion and kidney retention levels of oxalate, phosphate and calcium were significantly lowered upon supplementation with the extract in ethylene glycol induced rats. The alcoholic extract further reduced high serum levels of cretinine, urea nitrogen and uric acid as compared to standard cystone drug. Reduction in constituents forming the kidney stone and their decreased retention by the kidney reduces solubility product of crystallization salts including calcium phosphate and calcium oxalate that contribute a role in antiurolithiatic activity of the extract. Diuretic activity of the alcoholic extract at high doses was evidenced by increase in urine volume and concentration of Na^+ and K^+ [56].

11.17. *Raphanus Sativus*

R.sativus, popularly known as rabano negro (Radish) belonging to Cruciferae family is a common plant of forming an important part of cuisines worldwide especially Chinese, Korean and Japanese [57, 58]. This plant possesses significant antifungal activity. Apart from this, its leaves are as diuretic, laxative while its roots are employed for gastrodynic pains, urinary complications, haemorrhoids, and gastric ailments. The seeds on the other hand are expectorant, diuretic, digestive, laxative and carminative [57, 59]. The aqueous extract of the bark of *R.sativus* inhibited the urolith formation significantly in a dose dependent manner. Considerable decrease in stone weight on treatment with its aqueous extract was observed. An increase in the 24 hour urine on treatment with this extract was also observed. No significant reduction in body weight was observed [57].

11.18. *Rubia Cordifolia*

R.cordifolia belonging to the family Rubiaceae is widely distributed among the West Bengal and Uttarakhand in India commonly cultivated for its red coloured pigment. *R.cordifolia* is proven effective against cancer, Alzheimer, inflammation, hepatoprotective, nephroprotective, analgesic, immunomodulatory, enterocolitis, allergy etc. [60, 61]. Rats with Ethylene glycol induced urolithiasis were subjected to hydro - alcoholic extracts of roots of *R.cordifolia* to analyse its protective effect as well as the possible underlying mechanisms. Change in urinary calcium, phosphate and oxalate excretion was significantly inhibited on dose-dependent HARC supplementation. A significant revert in the number of calcium oxalate crystals deposits in the kidney tissue of calculogenic rats and increased calcium and oxalate levels was observed on HARC treatment. Impairment in renal functions after HARC supplementation was also observed. It was therefore, observed that HARC treatment prevented recurrence of the disease. The underlying mechanism may be conciliated possibly through nephroprotection, antioxidant and its effect on urinary concentration of stone forming constituents and risk factors [61].

11.19. *Viburnum Opulus L*

V.opulus belonging to the family Caprifoliaceae; the plant stands a native of Europe however spreading to Central zone of western Russia along with North Asia and North Africa. The medicinal properties of

V.opulus include its use in tuberculosis, cold and cough, kidney and bladder affections, duodenal ulcers, breath shortness, stomach and menstrual cramps [62, 63, 64]. The potential antiurolithiatic activity of Lyophilized juice of *V.opulus* (LJVO) and lyophilized commercial juice of *V.opulus* (LCJVO) stands responsible for its diuretic effects along with the inhibitory actions on free radical production and oxalate levels. Apart from this, no crystal deposition on sodium oxalate induced rats was observed. Increase in magnesium secretion in urine on treatment with LJVO and LCJVO was also observed. Apart from this significant increased levels of GSH and TSH were also observed. Decrease in serum creatinine, uric acid and urine nitrogen levels alond with glomerular filtration rate were seen on treatment with LJVO and LCJVO. Elevated levels of lipid peroxidation were also inhibited by LJVO and LCJVO [62].

11.20. *Xanthium Strumarium*

X.strumarium belonging to the family Asteraceae is a traditional medicinal plant found in India, Europe, China, America and Malaysia. The medicinal properties of the plant especially the roots, leaves and bur possess antidiabetic, anti-cell proliferative, antiulcerogenic, anti-inflammatory, analgesic, hypoglycaemic, antitrypanosomal and antihelmintic. Aqueous-ethanol extract of *X.strumarium* was subjected to urine-serum biochemistry, histopathology, oxidative/nitrosative stress indices, kidney calcium and calcium oxalate content and immunohistochemical expression of matrix glycoprotein, Osteopontin to evaluate its antiurolithiatic activity. X.strumarium administration significantly impaired hyperoxalurea, hypocalciurea, crystalluria, polyurea, increased serum creatinin, urea, erythrocytic lipid peroxidise and nitric oxide, kidney cacium content and crystal deposition along with decrease in up regulation of OPN [65].

Table 1. Medicinal plants as antiurolithiatic agents

S. No.	Plant Name	Family	Part of the plant analysed/used	Type of extract	Parameters analysed	References
1	*Abelmoschus moschatus*	Malvaceae	Seed	Chloroform and Methanol extract	Zinc disc induced urolithiatic rats.	30
2	*Acorus calamus*	Araceae	Rhizome	Ethanolic extract	Ethylene glycol induced urolithiatic rats.	31, 32
3	*Aerva lanata*	Amaranthaceae	Whole plant	(a) Ethyl acetate (b) n-butanol	Ethylene glycol induced urolithiatic rats	33, 34
4	*Betula alba*	Betulaceae	Betulin	DMSO (Dimethyl sulphoxide)	*In vitro* (ethylene glycol induced urolithiatic rats) and *In Silico* analysis	35
5	*Biophytum sensitivum*	Oxalidaceae	Whole plant	Ethyl acetate	Sodium oxalate induced urolithiasis	36, 37
6	*Carica papaya*	Caricaceae	Fruit	Aqueous and alcoholic extracts	Ethylene glycol induced urolithiatic rats	38, 39
7	*Cissampelos pareira*	Menispermaceae	Roots	Alcoholic extract	Ammonium chloride and ethylene glycol induced rats	40
8	*Costus spiralis*	Zingiberaceae	Whole plant	Hydro extract	Zinc disc implanted urolithiatic rats	41, 42
9	*Crataeva magna*	Capparaceae	Bark	Ethanol extract	Lactose+ Ethylene Glycol and Ammonium chloride+ Ethylene glycol induced urolithiatic rats	43, 44
10	Gokshuradi Polyherbal Formulation composed of: (a) *Tribulus terrestris*. (b) *Ricinnus communis L.* (c) *Solanum anguivi Lamk.* (d) *Solanum surattense Burm* (e) *Hygrophyta spinosa T.Anders*		(a) Dry fruit (b) Roots (c) Roots (d) Roots (e) Seeds	Hydro- extract	Ethylene Glycol induced urolithiatic rats.	45

S. No.	Plant Name	Family	Part of the plant analysed/used	Type of extract	Parameters analysed	References
11	*Holarrhena antidysenterica*	Acanthaceae	Whole plant	Aqueous extract	(a) MDCK Cell lines (b) Ethylene glycol induced urolithiatic rats	46, 47, 48, 66, 67, 68
12	*Hygrophila auriculata*	Apocynaceae	Whole plant	Alcoholic (n-butanol) extract	Ethylene glycol induced urolithiatic rats	49, 50
13	*Lanata camara*	Verbenaceae	Roots	(a) Ethanolic extract (b) oleanolic extract	Zinc disc induced urolithiatic rats.	51, 52
14	*Lepidagathis prostrate*	Acanthaceae	Whole plant	(a) Ethyl acetate (LPEA) (b) n-butanol	Sodium oxalate	9
15	*Paronychia argentea*	Illecebraceae	Aerial parts	Butanolic extract (BPA)	Sodium oxalate induced urolithiatic rats.	53, 54, 55
16	*Pergularia daemia*	Asclepiadaceae	Whole plant	Alcoholic extract	Ethylene glycol induced urolithiatic rats.	56
17	*Raphanus sativus*	Cruciferae	Bark	Hydro-extract	Zinc disc implanted urolithiatic rats.	57, 58, 59
18	*Rubia cordifolia*	Rubiaceae	Roots	Hydroalcoholic extract	Ethylene glycol induced urolithiatic rats.	60, 61,
19	*Viburnum opulus*	Caprifoliaceae	Fruits	(a) Lyophilized commercial juice of V.opulus (LCVJO) (b) Lyophilized juice of V.opulus (LJVO)	Sodium oxalate induced urolithiasis	62, 63, 64
20	*Xanthium strumarium*	Asteraceae	Bur	Hydroethanolic extract	Ethylene glycol and ammonium chloride induced urolithiatic rats	65

CONCLUSION

Urolithiasis is a condition occurring due to restricted excretion of stone forming components including calcium, oxalate, phosphate along with other factors. Urolithiasis, has a prevalence rate of 1-15% the variation of which depends on certain factors including age, gender, race, ethnic and familial backgrounds, lifestyle and dietary habits, climate and environment and educational and occupational levels. Certain components that may skip excretion through urine stand responsible for/act as main components of the calculi formed. Apart from them there are other body secretions that that may promote or inhibit calculi formation. ROS on the other hand defend the body against such calculi but may also prove to be detrimental if present in higher concentration as they may cause damage to healthy cells which again is hazardous. To overcome this drawback, drugs possessing antioxidant properties have been derived. The drugs derived from medicinal plants are found to possess antiurolithiatic as well as antioxidant activity which accounts for its advantage with minimised side-effects along with reduced stone recurrence rate. These drugs in combination with certain surgical techniques may prove fruitful in calculi remedification.

REFERENCES

[1] Barreto L, Jung JH, Abdelrahim A, Ahmed M, Dawkins GP, Kazmierski M. Medical and surgical interventions for the treatment of urinary stones in children. *Cochrane Database of Systematic Reviews.* 2018(6).

[2] Mikawlrawng K, Kumar S, Vandana R. Current scenario of urolithiasis and the use of medicinal plants as antiurolithiatic agents in Manipur (North East India): a review. *International Journal of Herbal Medicine.* 2014;2(1):1-2.

[3] Khan A, Bashir S, Khan SR, Gilani AH. Antiurolithic activity of Origanum vulgare is mediated through multiple pathways. *BMC complementary and alternative medicine.* 2011 Dec;11(1):96.

[4] Alelign T, Petros B. Kidney stone disease: an update on current concepts. *Advances in urology.* 2018;2018.

[5] Bawari S, Sah AN, Tewari D. Anticalcifying effect of Daucus carota in experimental urolithiasis in Wistar rats. *Journal of Ayurveda and integrative medicine.* 2019 Apr 5.

[6] El-Zoghby ZM, Lieske JC, Foley RN, Bergstralh EJ, Li X, Melton LJ, Krambeck AE, Rule AD. Urolithiasis and the risk of ESRD. *Clinical Journal of the American Society of Nephrology.* 2012 Sep 1;7(9):1409-15.

[7] Liu Y, Chen Y, Liao B, Luo D, Wang K, Li H, Zeng G. Epidemiology of urolithiasis in Asia. *Asian journal of urology.* 2018 Sep 6.

[8] Romero V, Akpinar H, Assimos DG. Kidney stones: a global picture of prevalence, incidence, and associated risk factors. *Reviews in urology.* 2010;12(2-3):e86.

[9] Devkar RA, Chaudhary S, Adepu S, Xavier SK, Chandrashekar KS, Setty MM. Evaluation of antiurolithiatic and antioxidant potential of *Lepidagathis prostrata*: a Pashanbhed plant. *Pharmaceutical biology.* 2016 Jul 2;54(7):1237-45.

[10] Ahmed S, Hasan MM, Mahmood ZA. Antiurolithiatic plants: Multidimensional pharmacology. *Journal of Pharmacognosy and Phytochemistry.* 2016 Mar 1;5(2):4.

[11] Nagal A, Singla RK. Herbal resources with antiurolithiatic effects: a review. *Indo Glob J Pharm Sci.* 2013;3(1):6-14.

[12] Verkoelen CF. Crystal retention in renal stone disease: a crucial role for the glycosaminoglycan hyaluronan?. *Journal of the American Society of Nephrology.* 2006 Jun 1;17(6):1673-87.

[13] Hirose M, Tozawa K, Okada A, Hamamoto S, Higashibata Y, Gao B, Hayashi Y, Shimizu H, Kubota Y, Yasui T, Kohri K. Role of osteopontin in early phase of renal crystal formation: immunohistochemical and microstructural comparisons with

osteopontin knock-out mice. *Urological research.* 2012 Apr 1;40(2):121-9.

[14] Gupta M, Bhayana S, Sikka SK. Role of urinary inhibitors and promoters in calcium oxalate crystallisation. *Int J Research in Pharmacy and Chemistry.* 2011;1:793-8.

[15] Basavaraj DR, Biyani CS, Browning AJ, Cartledge JJ. *The role of urinary kidney stone inhibitors and promoters in the pathogenesis of calcium containing renal stones.* EAU-EBU update series. 2007 Jun 1;5(3):126-36.

[16] Ryall RL. Glycosaminoglycans, proteins, and stone formation: adult themes and child's play. *Pediatric Nephrology.* 1996 Oct 1;10(5):656-66.

[17] Aggarwal KP, Narula S, Kakkar M, Tandon C. Nephrolithiasis: molecular mechanism of renal stone formation and the critical role played by modulators. *BioMed research international.* 2013;2013.

[18] Tsujihata M, Miyake O, Yoshimura K, Kakimoto KI, Takahara S, Okuyama A. Comparison of fibronectin content in urinary macromolecules between normal subjects and recurrent stone formers. *European urology.* 2001;40(4):458-62.

[19] Tsujihata M, Miyake O, Yoshimura K, Tsujikawa K, Tei N, Okuyama A. Renal tubular cell injury and fibronectin. *Urological research.* 2003 Dec 1;31(6):368-73.

[20] Xing J, Qin J, Cai Z, Duan B, Bai P. Association between calcitonin receptor gene polymorphisms and calcium stone urolithiasis: A meta-analysis. *International braz journal.* 2019 Sep;45(5):901-9.

[21] Basavaraj DR, Biyani CS, Browning AJ, Cartledge JJ. *The role of urinary kidney stone inhibitors and promoters in the pathogenesis of calcium containing renal stones.* EAU-EBU update series. 2007 Jun 1;5(3):126-36.

[22] Fakier S, Rodgers A, Jackson G. Potential thermodynamic and kinetic roles of phytate as an inhibitor of kidney stone formation: theoretical modelling and crystallization experiments. *Urolithiasis.* 2019 Dec 1;47(6):493-502.

[23] Hess B, Zipperle L, Jaeger P. Citrate and calcium effects on Tamm-Horsfall glycoprotein as a modifier of calcium oxalate crystal aggregation. *Am J Physiol.* 1993;265(6 Pt 2): F784-F791. doi:10.1152/ajprenal.1993.265.6.F784

[24] Flannigan R, Choy WH, Chew B, Lange D. Renal struvite stones—pathogenesis, microbiology, and management strategies. *Nature reviews Urology.* 2014 Jun;11(6):333.

[25] Rimer JD, An Z, Zhu Z, Lee MH, Goldfarb DS, Wesson JA, Ward MD. Crystal growth inhibitors for the prevention of L-cystine kidney stones through molecular design. *Science.* 2010 Oct 15;330(6002):337-41.

[26] Daudon M, Frochot V, Bazin D, Jungers P. Drug-induced kidney stones and crystalline nephropathy: pathophysiology, prevention and treatment. *Drugs.* 2018 Feb 1;78(2):163-201.

[27] Perazella MA. Drug-induced nephropathy: an update. *Expert opinion on drug safety.* 2005 Jul 1;4(4):689-706.

[28] Strakosha R, Monga M, Wong MY. The relevance of Randall's plaques. *Indian journal of urology: IJU: journal of the Urological Society of India.* 2014 Jan;30(1):49.

[29] Andrabi Y, Patino M, Das CJ, Eisner B, Sahani DV, Kambadakone A. Advances in CT imaging for urolithiasis. *Indian journal of urology: IJU: journal of the Urological.*

[30] Pawar AT, Vyawahare NS. Antiurolithiatic activity of Abelmoschus moschatus seed extracts against zinc disc implantation-induced urolithiasis in rats. *Journal of basic and clinical pharmacy.* 2016 Mar;7(2):32.

[31] Ghelani H, Chapala M, Jadav P. Diuretic and antiurolithiatic activities of an ethanolic extract of Acorus calamus L. rhizome in experimental animal models. *Journal of traditional and complementary medicine.* 2016 Oct 1;6(4):431-6.

[32] Subrahmanyam TV. Acorus calamus-the Sweet-flag-a new indigenous Insecticide for the Household. *Indian Journal of Entomology.* 1942;4(pt. 2).

[33] Dinnimath BM, Jalalpure SS, Patil UK. Antiurolithiatic activity of natural constituents isolated from Aerva lanata. *Journal of Ayurveda and integrative medicine.* 2017 Oct 1;8(4):226-32.

[34] Goyal M, Pareek A, Nagori BP, Sasmal D. Aerva lanata: A review on phytochemistry and pharmacological aspects. *Pharmacognosy reviews.* 2011 Jul;5(10):195.

[35] Nirala RK, Dutta P, Malik MZ, Dwivedi L, Shrivastav TG, Thakur SC. In Vitro and In Silico Evaluation of Betulin on Calcium Oxalate Crystal Formation. *Journal of the American College of Nutrition.* 2019 Oct 3;38(7):586-96.

[36] Pawar AT, Vyawahare NS. Protective effect of ethyl acetate fraction of Biophytum sensitivum extract against sodium oxalate-induced urolithiasis in rats. *Journal of traditional and complementary medicine.* 2017 Oct 1;7(4):476-86.

[37] Abhirama BR. *In vitro and In vivo Studies on Nephroprotective and Antiurolithic Effect of Biophytum sensitivum in Different Toxicity Models* (Doctoral dissertation, JKK Nattraja College of Pharmacy, Komarapalayam).

[38] Nayeem K, Guptaa D, Nayanab H, Joshic RK. http://jprsolutions. info. *Journal of Pharmacy Research.* 2010 Nov;3(11):2772-5.

[39] Singh SP, Rao DS. Papaya (Carica papaya L.). In *Postharvest biology and technology of tropical and subtropical fruits* 2011 Jan 1 (pp. 86-126e). Woodhead Publishing.

[40] Babu Sayana SU, Khanwelkar CC, Rao Nimmagadda VE, Chavan VR, BH R, Kumar S NA. Evaluation of Antiurolithic Activity of Alcoholic Extract of Roots of Cissampelos Pareira in Albino Rats. *Journal of Clinical & Diagnostic Research.* 2014 Jul 1;8(7).

[41] Viel TA, Domingos CD, da Silva Monteiro AP, Lima-Landman MT, Lapa AJ, Souccar C. Evaluation of the antiurolithiatic activity of the extract of Costus spiralis Roscoe in rats. *Journal of ethnopharmacology.* 1999 Aug 1;66(2):193-8. [*C. spiralis*].

[42] Calaboni C, Martins MB, Rossi ML. Anatomical characterization of *Costus spiralis* (Jacq.) Roscoe leaves from impacted and non-

impacted environments from the São Paulo coast. Iheringia, *Série Botânica*. 2013;68(2):225-35. [*C. spiralis*].

[43] Mekap SK, Mishra S, Sahoo S, Panda PK. Antiurolithiatic activity of *Crataeva magna* Lour. bark.

[44] Nagarajan K. Isolation and Characterisation of Crataeva Magna Lour (DC) Ethanolic Extract and its Anti Cancer Activity Against Dalton's Ascites Lymphoma Cell Line (Doctoral dissertation, KM College of Pharmacy, Madurai). [*Crataeva magna* Lour].

[45] Shirfule AL, Racharla V, Qadri SS, Khandare AL. Exploring antiurolithic effects of gokshuradi polyherbal ayurvedic formulation in ethylene-glycol-induced urolithic rats. *Evidence-Based Complementary and Alternative Medicine*. 2013;2013. [Gokshuradi].

[46] Khan A, Khan SR, Gilani AH. Studies on the in vitro and in vivo antiurolithic activity of Holarrhena antidysenterica. *Urological research*. 2012 Dec 1;40(6):671-81.

[47] Sharma DK, Gupta VK, Kumar S, Joshi V, Mandal RS, Prakash AB, Singh M. Evaluation of antidiarrheal activity of ethanolic extract of Holarrhena antidysenterica seeds in rats. *Veterinary world*. 2015 Dec;8(12):1392.

[48] Dwivedi RK, Tripathi YC. Pharmacognostical, phytochemical and biological studies on Holarrhena antidysenterica Wall: a review (part-1). *New Agriculturist*. 1991;1(2):209-12.

[49] Sethiya NK, Ahmed NM, Shekh RM, Kumar V, Singh PK, Kumar V. Ethnomedicinal, phytochemical and pharmacological updates on Hygrophila auriculata (Schum.) Hiene: an overview. *Journal of integrative medicine*. 2018 Sep 1;16(5):299-311.

[50] Hussain S, Ahmed N, Ansari Z. Preliminary studies on diuretic effect of Hygrophila auriculata (Schum) Heine in rats. *International Journal of Health Research*. 2009;2(1).

[51] Vyas N, Argal A. Antiurolithiatic activity of extract and oleanolic acid isolated from the roots of Lantana camara on zinc disc implantation induced urolithiasis. *ISRN pharmacology*. 2013 May 15;2013.

[52] Lonare MK, Sharma M, Hajare SW, Borekar VI. Lantana camara: overview on toxic to potent medicinal properties. *International Journal of Pharmaceutical Sciences and Research.* 2012 Sep 1;3(9):3031.

[53] Bouanani S, Henchiri C, Migianu-Griffoni E, Aouf N, Lecouvey M. Pharmacological and toxicological effects of Paronychia argentea in experimental calcium oxalate nephrolithiasis in rats. *Journal of ethnopharmacology.* 2010 May 4;129(1):38-45.

[54] Zama D, Meraihi Z, Tebibel S, Benayssa W, Benayache F, Benayache S, Vlietinck AJ. Chlorpyrifos-induced oxidative stress and tissue damage in the liver, kidney, brain and fetus in pregnant rats: The protective role of the butanolic extract of Paronychia argentea L. *Indian Journal of Pharmacology.* 2007 May 1;39(3):145.

[55] Belarbi Z, Gamby J, Makhloufi L, Sotta B, Tribollet B. Inhibition of calcium carbonate precipitation by aqueous extract of Paronychia argentea. *Journal of crystal growth.* 2014 Jan 15;386:208-14.

[56] Vyas BA, Vyas RB, Joshi SV, Santani DD. Antiurolithiatic activity of whole-plant hydroalcoholic extract of Pergularia daemia in rats. *Journal of young pharmacists.* 2011 Jan 1;3(1):36-40.

[57] Vargas S R, Perez G RM, Perez G S, Zavala S MA, Perez G C. Antiurolithiatic activity of Raphanus sativus aqueous extract on rats. *Journal of ethnopharmacology.* 1999 Dec 15;68(1-3):335-8.

[58] Wang N, Kitamoto N, Ohsawa R, Fujimura T. Genetic diversity of radish (Raphanus sativus) germplasms and relationships among worldwide accessions analyzed with AFLP markers. *Breeding science.* 2008;58(2):107-12.

[59] Alqasoumi S, Al-Yahya M, Al-Howiriny TA, Rafatullah SY. Gastroprotective effect of radish "Raphanus sativus'" L. on experimental gastric ulcer models in rats. Farmacia-Bucuresti. 2008;56(2):204.

[60] Divakar K, Pawar AT, Chandrasekhar SB, Dighe SB, Divakar G. Protective effect of the hydro-alcoholic extract of Rubia cordifolia roots against ethylene glycol induced urolithiasis in rats. *Food and Chemical Toxicology.* 2010 Apr 1;48(4):1013-8.

[61] Bhatt P, Kushwah AS. Rubia cordifolia overview: A new approach to treat cardiac disorders. *Int. J. Drug Dev. & Res.* 2013 Apr;5(2):47-54.

[62] İlhan M, Ergene B, Süntar I, Özbilgin S, Saltan Çitoğlu G, Demirel MA, Keleş H, Altun L, Küpeli Akkol E. Preclinical evaluation of antiurolithiatic activity of Viburnum opulus L. on sodium oxalate-induced urolithiasis rat model. *Evidence-Based Complementary and Alternative Medicine.* 2014;2014.

[63] Sagdic O, Aksoy A, Ozkan G. Evaluation of the antibacterial and antioxidant potentials of cranberry (gilaburu, Viburnum opulus L.) fruit extract. *Acta Alimentaria.* 2006 Dec 1;35(4):487-92.

[64] Sedat Velioglu Y, Ekici L, Poyrazoglu ES. Phenolic composition of European cranberrybush (Viburnum opulus L.) berries and astringency removal of its commercial juice. *International journal of food science & technology.* 2006 Nov;41(9):1011-5.

[65] Panigrahi PN, Dey S, Sahoo M, Choudhary SS, Mahajan S. Alteration in Oxidative/nitrosative imbalance, histochemical expression of osteopontin and antiurolithiatic efficacy of Xanthium strumarium (L.) in ethylene glycol induced urolithiasis. *Biomedicine & Pharmacotherapy.* 2016 Dec 1;84:1524-32.

[66] De Kruif AP. *A revision of Holarrhena R. Br. (Apocynaceae).* Veenman; 1981.

[67] Caius JF, Mhaskar KS. A Study of Indian Medicinal Plants. Holarrhena antidysenterica, Wall. *A Study of Indian Medicinal Plants. Holarrhena antidysenterica*, Wall. 1927.

[68] Waugh A, Grant A. *Ross and Wilson Anatomy and physiology in health and illness.* London: Churchill Livingstone; 2001 Aug.

In: Oxalate
Editor: Elsa Kytönen

ISBN: 978-1-53618-303-0
© 2020 Nova Science Publishers, Inc.

Chapter 2

TRANSITION METAL OXALATES AS POTENTIAL FUTURISTIC MATERIALS FOR EFFICIENT ENERGY STORAGE CAPACITY

Sushree Pattnaik, Arya Das, Kishor Kumar Sahu, Suddhasatwa Basu and Mamata Mohapatra[*]

CSIR-Institute of Minerals and Materials Technology, Bhubaneswar- Odisha, India

ABSTRACT

Transition Metal oxalate is one of the pragmaticmaterial forcatalytic, sensor, battery and SCs application due to its low cost, environmental friendliness, and good chemical stability. In this regard a lot of research work has recently focused on implementation of it as potential electrode materials for enhanced energy storage. Owing to their unique property to act as a carbon sink further establishes them as sustainable and green materials for energy storage, thus strengthening their stand among variety

---

[*]Corresponding Author's Email:mamata@immt.res.in.

of materials available for possible commercialization. This progress report presents the various known transition metal oxalates, their synthetic analogues, thermal and magnetic behaviour, the reaction pathways, particularly for energy materials such as Li ion, supercapacitor and redox flow battery systems. Finally, the challenges of these materials in these energy storge systems for high power and energy output are outlined.

Keywords: transition metal oxalates, thermal and magnetic properties, energy storage, lithium ion battery

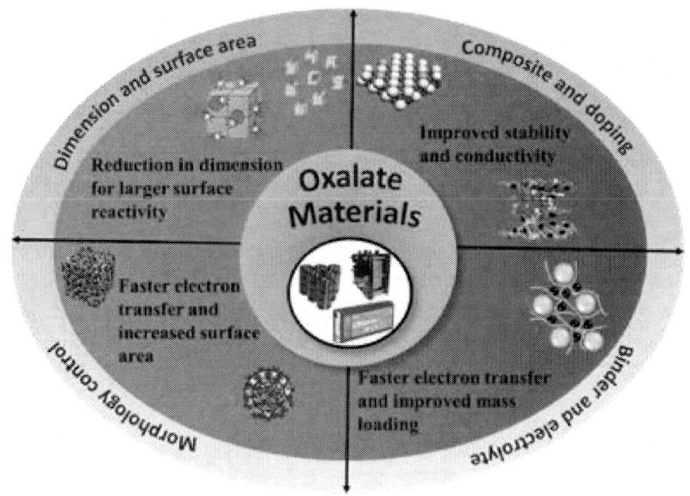

INTRODUCTION

Organic minerals constitute class 10 in the mineralogical classification [1]. The classification includes simple and complex salts of the organic acids like formic, citric, acetic, mellitic, methane sulfonic and oxalic acid shown in Figure 1. Two natural formates, three natural acetates, one natural citrate, one mellite, one salt of methane sulfonic acid and 21 natural oxalate minerals have been so far reported and studied. Oxalates constitute the most abundant class of organic minerals distributed in nature [2]. Oxalic acid ($C_2H_2O_4$) is also referred as ethanedioic acid and is the simplest dicarboxylic acid. Its dianion $C_2O_4^{2-}$ is obtained by deprotonation of both carbon groups of $C_2H_2O_4$. Figure 2 shows structure of an oxalate ion. A wide range of salts and complexes of oxalates has been reported. Though it's a tetradentate ligand, but it also acts as mono, bi- or tridentate depending on the requirement of forming complexes [3].

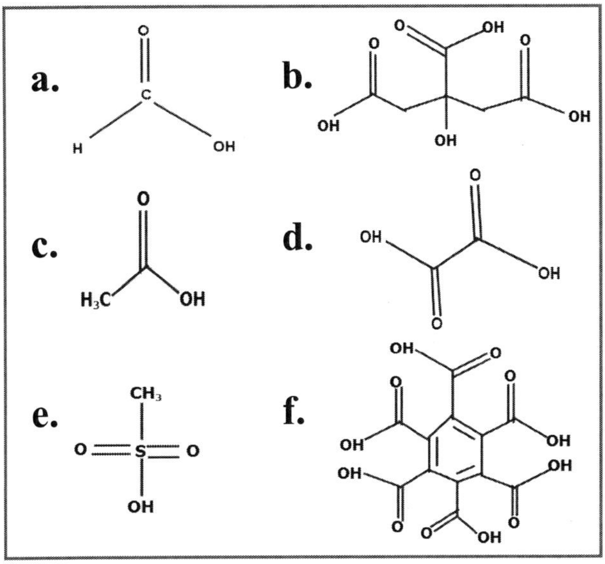

Figure 1. Organic acids (a) formic acid (b) citric acid (c) acetic acid (d) oxalic acid (e) methanesulfonic acid and (f) mellitic acid whose organic minerals are reported and studied.

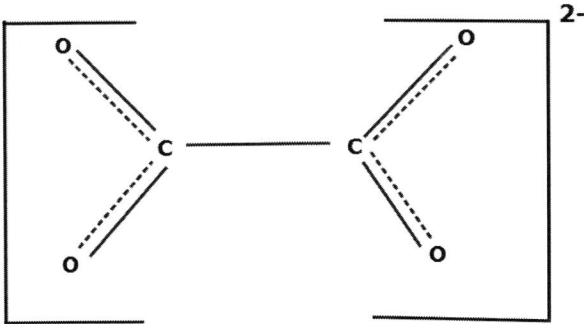

Figure 2. Structure of oxalate ion.

TRANSITION METAL OXALATES

Many first row transition elements have been reported of forming oxalate complexes. They generally form dihydrate oxalates with α and β modifications [6, 7, 8]. These complexes reported to be having two molecules of water at each end of cation giving an octahedron around the metal [5, 6, 7, 8, 9].

The chains form sheets that give α and β modifications due to the difference in stacking causing the reshuffling of the H-bonds present [8]. The α modification is found to be more thermodynamically stable than β and its conversion from latter to former is not reversible [6]. Both the forms are reported to be obtained via synthetic procedures [6, 7]. The α form is reported to be synthesized using soluble salt of cation with slight excess of oxalic acid or alkaline oxalate and β form is obtained under similar conditions using a defect of oxalic acid and solution being boiled during course of the reaction. Structures of both forms were proposed based on XRD [6, 7, 8, 10, 11].

Mn (II) reports both the α and β forms along with γ-Mn $(C_2O_4) \cdot 2H_2O$ and Mn $(C_2O_4) \cdot 3H_2O$ [40]. Only α form has been reported so far for Zn (II) [7, 8]. The α form has been reported to be obtained at room temperature for Mn(II), Fe(II) and Zn(II) whereas Co(II) is reported to be requiring a temperature of 50°-60°C and Ni(II) requires heating under reflux [6, 7, 8, 10, 11]. Cu reports its oxalate Cu $(C_2O_4) \cdot nH_2O$ obtained by reaction of oxalic acid with $CuSO_4 \cdot 5H_2O$ [13]. A ribbon like structure similar to as shown in Figure 3 has been reported. A square planar configuration with Cu-O distance of 1.98(2) Å and Cu-Cu distance of 5.14 Å has been reported for Cu (II) [14]. It has been reported for Mn (II) its trihydrate is obtained by reacting dilute Mn (II) acetate solution to other with stoichiometric oxalic acid.

Upon complete addition a pink precipitate is reported to be formed and its boiling leads to forming a suspension generating α-Mn $(C_2O_4) \cdot 2H_2O$. It has also been reported that high pH-value favors formation of γ-Mn $(C_2O_4) \cdot 2H_2O$ by reacting dilute ammonium oxalate with the solution of

Mn(II) nitrate at a temperature of 80 °C [47]. Mn (II) in its distorted octahedral form is reported to have two molecules of water in trans and four of oxygen one non-bonded and other bonded to both the Mn cations. [15]. Two of the research groups reports the structure of the trihydrate as [Mn C_2O_4 (H_2O)$_2$] · H_2O [16, 17]. Mn (II) is reported to be bridged bis-bidentately by oxalate groups creating a zig-zag chain in 1D. Mn (II) ions reportedly coordinates with two third of molecules of water present and rest remains as lattice water. The two molecules of water are in cis to each other. When dry trihydrate is reported to be unstable and its conversion to a form takes few days. Mn in its anhydrous oxalate form is obtained via vacuo heating [12]. Figure 4 shows structure of [Mn C_2O_4 (H_2O)$_2$]·H_2O. Mn (II) as [Mn_2 (C_2O_4)(OH)$_2$] reported to have a 3D complex structure is formed reacting hydrothermally mixtures of MnC_2O_4·$2H_2O$, piperazine and water at 160 °C [15]. Cobalt oxalate tetrahydrate has been investigated. It has been reported to be obtained by diffusing solution 4-amine-1,2,4-triazole in methanol with Co(C_2O_4)·$2H_2O$ and K_2(C_2O_4)·H_2O and left for a span of two weeks to get red crystals of desired oxalate of Co [18]. Amino acid L- serine is used as the source of carboxylate to obtain Ni (II) oxalate dihydrate via solvothermal method. Its reports a distorted octahedral structure containing a total of 6 oxygen atoms where four are contributed from two different oxalates and remaining two from water molecules [19]. Both Fe (II) and Mn (II) oxalates are reported to be isostructural. Figure 5 shows Infinite chain structure in humboldtine and other divalent metal cation oxalates of type M^{II} (C_2O_4)·$2H_2O$. Humboldtine decomposes to give $Fe^{(II)}$ (C_2O_4)·$2H_2O$ complex [20]. Trihydrate of Cd oxalate is reported to be obtained as a precipitate by reacting Cd (II) acetate and oxalic acid at 65 °C [21]. Silica gel saturated with oxalic acid produces single crystals of Cd(II) [22] or in situ method with oxalic acid and aqueous Cd(II) solutions, oxalylhydroxamic acid [23] or L-ascorbic acid [24]. For Pb trihydrate [23] along with dihydrate [53] and anhydrous [26, 27] oxalate complexes has been studied. A variety of oxalato complexes with chains of metal-oxalate along with a number of secondary ligands have been reported. In the Cu oxalate complex with hydroxyppyri-dine each of the cation is coordinated with four oxygen from two of the bridged oxalato ligands and two of the nitrogen atoms from the two pyOH molecule. The chain has a zig-zag structure with two of the ligands in cis position [28]. [$Cu_2(C_2O_4)_2$(ampy)$_3$]·ampy (ampy = 2-amino-4-methylpyridine) reports a complex structure.

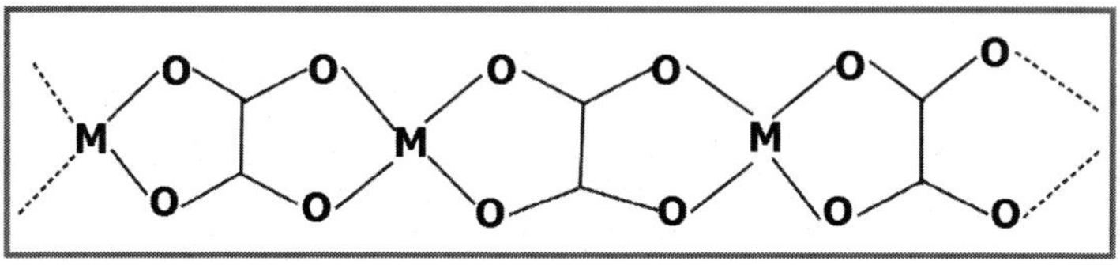

Figure 3. Infinite chain structure in divalent metal cation oxalates of type M^{II} (C_2O_4)·$2H_2O$.

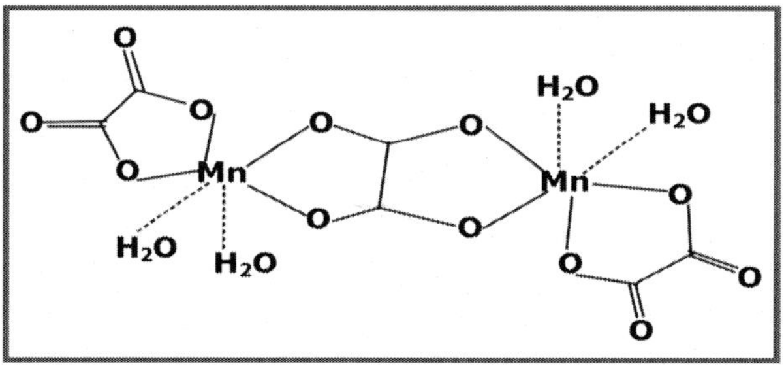

Figure 4. Mn (II) environment in [Mn C_2O_4 (H_2O)$_2$]·H_2O.

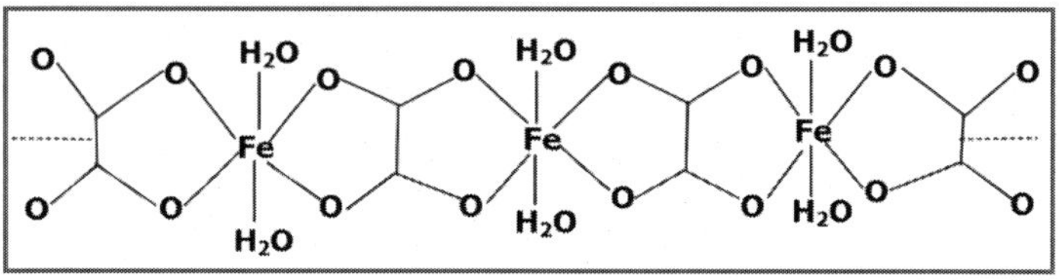

Figure 5. Infinite chain structure in humboldtine and other divalent metal cation oxalates of type M^{II} (C_2O_4)·$2H_2O$.

Its structure reportedly comprises of ampy molecules, Cu (II) ions [Cu(1) and Cu(2)] chains connected via bis (chelating) oxalato ligands and two independent Cu (II) ions swapping frequently within the chain. An extended octahedral environment is made around Cu (1) using four of the oxygen atoms of bridging oxalate anions and two N-atoms from ampy ligands which were cis-coordinated. Cu (2) ion reports a distorted square pyramid structure where its basal plane comprises of four oxalate–oxygen atoms and apical plane has pyridine nitrogen atom from ampy molecule. The reaction of complex $M(C_2O_4) \cdot 2H_2O$ (M = Co(II), Ni(II)) or $K_2[Cu(C_2O_4)_2] \cdot 2H_2O$ with n-ampy (n = 2,3,4; n-ampy = n-aminopyridine) and oxalate of potassium reports a remarkable series of connected complexes. Generally, it reports formation of 1-D oxalato-bridged metal (II) complexes where its coordination sphere is concluded using a pair of aromatic bases [29]. Hydrothermally synthesized two similar complexes of Fe and Ni Fe $(C_2O_4) \cdot DPA$ and $Ni(C_2O_4) \cdot DPA$ (DPA = 2,2-dipyridylamine) are obtained [30]. From 1D zig-zag chains of oxalates fascinating complexes of $Zn(C_2O_4) \cdot bipy$ [31], $Co(C_2O_4) \cdot bipy$ [32] (bipy = 2,2′-bipyridine), and $Cu(C_2O_4) \cdot ophen$ (ophen = 1,10-phenanthroline) [33] has been reported. It has been stated that Fe(II), Co(II), Ni(II), Zn(II), and Cu(II) when thermally decomposed gives compounds which are anhydrous in two modifications first α-MC_2O_4 which is disordered and second β-MC_2O_4 which is ordered [34]. The suspension of Cu $(C_2O_4) \cdot nH_2O$ made in warm solution of its respective monovalent cation oxalate with a molar ratio of soluble oxalate/Cu $(C_2O_4) \cdot nH_2O$ equal to 2 gives $Na_2[Cu(C_2O_4)_2] \cdot 2H_2O$, and the analogous potassium and ammonium compounds [35]. $Na_2[Cu(C_2O_4)_2] \cdot 2H_2O$ comprises of quasi-planar units of $[Cu(C_2O_4)_2]^{2-}$ and linear chains of Na^+ cations and H_2O molecules which is parallel to c-axis of unit cell. A square CuO_4 structure has been reported with Cu (II) lying on the centre coordinated with oxalate anions, acting as bidentate ligand. The coordination sphere of copper is completed using two oxygen from $[Cu(C_2O_4)_2]^{2-}$ groups which is centro-symmetrically related each at both ends in the same pile [36]. An intricate process has been reported for synthesis of $Rb_2[Cu(C_2O_4)_2] \cdot 2H_2O$ which includes slow evaporation of the aqueous solutions with Rb_2CO_3, $CuCl_2$, basic copper carbonate, $H_2C_2O_4 \cdot 2H_2O$, and HNO_3 [37]. The process for synthesis of $Cs_2[Cu(C_2O_4)_2] \cdot 2H_2O$ has not been elaborated [38], but it has been reported that it can be synthesized through similar route as that for rubidium compound. Two kinds of units are reported for $K_2[Cu(C_2O_4)_2] \cdot 2H_2O$ and of the isotypic NH_4^+, Rb^+, and Cs^+ salts which happens to be $[Cu(C_2O_4)_2(H_2O)_2]^{2-}$ and $[Cu(C_2O_4)_2]^{2-}$ groups in a way that Cu(II) ions interact using the weak bridges of oxalates which includes free O-atoms from ligands of oxalates in first unit. It concludes that in both the structures Cu (II) is positioned in a distorted CuO_6 octahedral coordination [36].

Oxalato Complexes Containing Fe (III)

An equivalent synthetic structure to minguzzite is a complex of Fe (III), which is K_3 [Fe $(C_2O_4)_3$]·$3H_2O$. It has been reported that this structure can simply be synthesized by different processes like using peroxide for Fe(II) oxalate oxidation in presence of potassium oxalate and oxalic acid [39] by reacting solutions of $FeCl_3$ and $K_2(C_2O_4) \cdot H_2O$ [40] and also by digesting the mixture of ferric sulfate, barium oxalate, and $K_2(C_2O_4) \cdot H_2O$ [41]. The complex K_3 [Fe$(C_2O_4)_3$]·$3H_2O$ is photosensitive, the ligand oxalate gets oxidized to CO_2 and also reduction of Fe(III)–Fe(II) [42]. The above mentioned structure has been reported to be isostructural with the complexes of Al (III), Cr (III), and Ru (III) complexes [43].

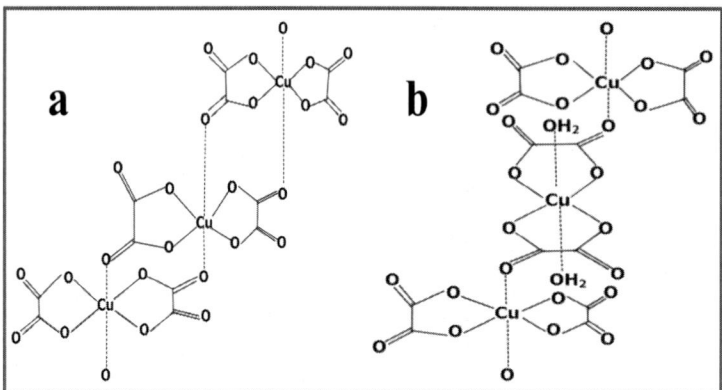

Figure 6.(a) Structural units in $Na_2 Cu(C_2O_4)_2 \cdot 2H_2O$ and (b) in the respective K, Rb, Cs, and NH_4 salts.

The complexes obtained via synthetic processes for stepanovite, Na Mg [$Fe^{III}(C_2O_4)_3$]·8–9H_2O, and zhemchuzhnikovite, Na Mg [(Al,Fe^{III})$(C_2O_4)_3$]·8H_2O has not been reported yet. The complex Na Mg

$[Al(C_2O_4)_3] \cdot 8H_2O$ is reported to be used in spectroscopic studies. It can be synthesized reacting $Al_2(SO_4)_3$, $Ba(OH)_2$, NaOH, MgO, and $H_2C_2O_4 \cdot 2H_2O$ in water in stoichiometric amounts. The partly replacement of Al (III) by other potential trivalent cations generates zhemchuzhnikovite, and its overall replacement gives stepanovite [44]. Detailed study on structural aspects has not been reported for these mentioned complexes yet. $Fe_2(C_2O_4)_3 \cdot 4H_2O$ is reported to be synthesized using ferric hydroxide with oxalic acid [45]. Presence of carbonyl group at the terminal position has been confirmed using FTIR and Raman hence it has excluded option of having an ionic structure. A dimeric structure has been expected based on stoichiometry. Figure 7 depicts a probable structural arrangement based on 57Fe-Mössbauer spectra that is reported to establish similarity between the dimeric positions and the octahedral coordination with one metal centre, disturbing the structure. According to Figure 4, it states the existence of a tetradentate ligand of oxalate at centre which bridges two Fe (III) cations and bidentate ligands of oxalate at terminal positions are reported to be in connection with one metal centre. Two molecules of water are reported to be present perpendicular to the iron/oxalate linkage completing the coordination [46].

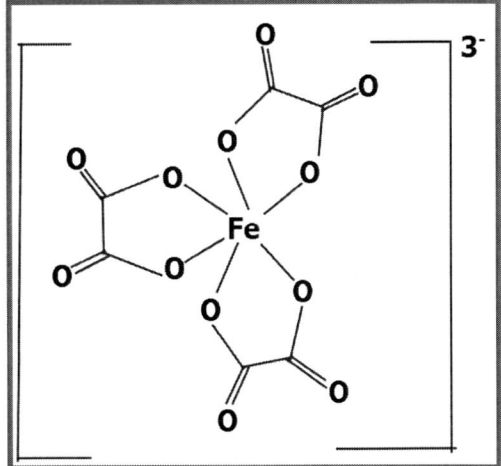

Figure 7. Fe (III) coordination sphere in $Fe[(C_2O_4)_3]^{3-}$.

Thermal Properties

For natural and synthetic oxalates reported so far their thermal properties under various parameters have been studied. There are three important reactions depicting the thermolysis process of divalent-metal oxalato complexes [5, 48, 49] as follows:

$$M(C_2O_4) \rightarrow MCO_3 + CO \tag{1}$$
$$M(C_2O_4) \rightarrow MO + CO + CO_2 \tag{2}$$
$$M(C_2O_4) \rightarrow M + 2CO_2 \tag{3}$$

Thermal Behavior of Natural Oxalates

The thermal properties of the natural oxalates were reported by R.L. Frost and his coworkers with the help of high resolution thermogravimetry in N_2 atmosphere also by evolving gas mass spectrometry, Raman microscopy and IR spectroscopy.

Naturally Occurring Transition Metal Oxalates

A complete thermal degradation of Cu (II) oxalate has been reported in two steps. Conversion of Cu oxalate to Cu oxide at 240 °C is reported to be occurring in the first step whereas generation of Cu_2O at 800°C takes place in the second step [50].

$$Cu(C_2O_4) \rightarrow CuO + CO_2 + CO \quad T=240°C \tag{4}$$

$$2CuO \rightarrow Cu_2O + \tfrac{1}{2}O_2 \quad T=800°C \tag{5}$$

High resolution thermogram so obtained for humboldtine that is Fe (II) oxalate reports complexity. For its thermal degradation three steps has been been reported which again constitutes multiple steps. At temperatures

130°C and 141°C two mass losses has been reported with a broad and not so defined step at 235°C in the first and second stage. Two mass losses at 312°C and 332°C have been reported in third stage. H_2O is reported to be liberated in two steps at first stage according to TG which has been confirmed using mass spectrometry. At a temperature of 200°C CO_2 begins to liberate which has been confirmed by mass spectra showing two peaks at 322°C and 354°C. A mechanism of decomposition has been proposed based on the information so obtained from Raman and IR spectra [51]:

$$Fe(C_2O_4) \cdot 2H_2O \rightarrow Fe(C_2O_4) \cdot H_2O + H_2O \qquad T=130°C \qquad (6)$$

$$Fe(C_2O_4) \cdot H_2O \rightarrow Fe(C_2O_4) + H_2O \qquad T=141°C \qquad (7)$$

Two alternate yet simultaneous ways has been reported for decomposition of anhydrous ferrous oxalate so obtained:

$$Fe(C_2O_4) \rightarrow FeO + CO + CO_2 \qquad T=200°C \qquad (8)$$

$$Fe(C_2O_4) \rightarrow FeCO_3 + CO \qquad T=322°C \qquad (9)$$

$$FeCO_3 \rightarrow FeO + CO_2 \qquad T=354°C \qquad (10)$$

Thermal Behavior of the Synthetic Transition Metal Oxalates

The thermal property of synthetic metal oxalates which also includes TG/DTA/DSC measurements under different parameters in isothermal conditions has been reported [48, 49, 52–54].

Transition metal oxalate of M^{II} cations has been reported. For both the natural and synthetic complexes of Fe^{II} thermal decomposition has been reported to be similar and complex. The atmosphere used for the degradation is an important parameter and various investigation were reported [54, 56, 57].

The thermal degradation of Fe^{II} oxalate reports formation of Fe_2O_3 a solid residue and Fe_3O_4 in minimal amount in N_2 atmosphere. When it is degraded in H_2 atmosphere elemental Fe is reported with minimal amount of Fe_2O_3 and Fe_3O_4. When treated in air Fe_2O_3 has been reported as the main product along with small amount of Fe_3O_4 and elemental Fe [54]. It has been reported thermal treatment of $Fe^{II}(C_2O_4).2H_2O$ gives gaseous products CO, CO_2 and H_2O, and solids were identified and characterized using ^{57}Fe Mossabauer spectroscopy and X-ray diffraction[58]. A five step mechanism of oxalate thermal degradation has been proposed:

Liberation of molecule of water
$$Fe^{II}C_2O_4.2H_2O \rightarrow Fe^{II}C_2O_4 + 2H_2O \qquad T=170\text{-}230°C \qquad (11)$$

Conversion of anhydrous oxalate to magnetite and Co
$$3Fe(C_2O_4) \rightarrow Fe_3O_4 + 4CO + 2CO_2 \qquad T=\text{above } 230°C \qquad (12)$$

Reduction of anhydrous oxalate by CO to Fe_3C
$$3Fe(C_2O_4) + 2CO \rightarrow Fe_3C + 7CO_2 \qquad T = \text{above } 360°C \qquad (13)$$

Conversion of carbide
$$Fe_3C \rightarrow 3Fe + C \qquad T=415\text{-}535°C \qquad (14)$$

Thermal reduction of magnetite into FeO by CO
$$Fe_3O_4 + CO \rightarrow 3FeO + CO_2 \qquad T=\text{above } 535°C \qquad (15)$$

As reported by the researchers decomposition process for Ferric oxalate is quite complex [59]. First step of dehydration reports overlap of both the liberation of CO_2 and ferrous oxalate decomposition resulting in a little amount of ferric oxalate which is undecomposed as residue [45]. The $\alpha\text{-}Fe_2O_3$ has been reported as the final product in oxidative conditions [60].

Inert atmosphere reports decomposition at a temperature 210°C generating Fe_3O_4.

$$Fe_2(C_2O_4)_3 \rightarrow 2Fe(C_2O_4) + 2CO_2 \qquad (16)$$

$$3Fe_2(C_2O_4)_3 \rightarrow 2Fe_3O_4 + 10CO_2 + 8CO \tag{17}$$

Dehydration process is reported to take place up to 300°C. Ferrous oxalate decomposition takes place in a range of 300-390°C.

$$Fe(C_2O_4) \rightarrow FeO + CO + CO_2 \tag{18}$$

Temperature below 570°C reports disproportionation of thermally unstable FeO.

$$4FeO \rightarrow Fe_3O_4 + Fe \tag{19}$$

Above the already mentioned temperature, syn proportionation takes place:

$$Fe + Fe_3O_4 \rightarrow 4FeO \tag{20}$$

Fe_2O_3 is reported as the ultimate product in oxidative conditions. Zn (II) oxalate decomposition has been reported to be simple. As per the observations so reported ZnO is obtained along with CO and CO_2 in all the atmospheres [54, 61]. For Ni $(C_2O_4).2H_2O$ degraded in air or N_2 atmosphere reports NiO as product, for He or H_2 atmosphere metallic Ni has been reported as the product [54, 62, 63]. For Co oxalate Co_3O_4 is reported to be the product in air and N_2 atmosphere and metallic Co in the H_2 atmosphere and also in case of inert atmospheres [54,64]. In case of Co oxalate it has been reported to release water molecules in two steps and in the range of 40-70°C lattice water is released and coordinated water at a range of 160-200°C. At a temperature of 280°C anhydrous oxalate of Co is reported to be stable and gives Co_3O_4 as product [18]. Thermal properties of α-Mn $(C_2O_4).2H_2O$ and Mn $(C_2O_4).3H_2O$ in air have been compared and many fascinating differences have been reported. For dihydrate dehydration process starts at 130°C. Lower temperature reports loss of water in case of trihydrate involving three steps [65]. Anhydrous oxalate of Mn when undergoes final decomposition reports the formation of Mn (III) and Mn (IV) species. It has also been reported that the crystal lattice stabilizes lower oxidation of Mn. Mn $(C_2O_4).2H_2O$ transforms into Mn_3O_4 when final annealing takes place at 450°C and also reported to give Mn_3O_4 along with Mn_2O_3 at the same temperature [65].

Cu (II) oxalate decomposition has been reported to be a complex one and is dependent on the atmosphere [54, 66, 67]. Zeolitic type water is reported to be lost between 40 and 185°C in air [66]. CuO is obtained as the final product [54, 67] along with Cu and Cu_2O as the intermediates. Ar or N_2 atmospheres reports metallic Cu with small amounts of Cu_2O and traces of CuO [54, 67]:

$$Cu (C_2O_4) \rightarrow Cu + 2CO_2 \tag{21}$$

$$2Cu (C_2O_4) \rightarrow Cu_2O + CO + 3CO_2 \tag{22}$$

$$Cu (C_2O_4) \rightarrow CuO + CO + CO_2 \tag{23}$$

Trihydrate of Cd oxalate reports dehydration in a single step at temperature of 31-179°C. At about 340°C $CdCO_3$ is reported to be generated and above 700°C reports CdO as the final product [22]. Wheatleyite has been reported to be studied among the series of complex oxalates [68]. Water molecule is reported to be eliminated at three consecutive steps at 78, 100 and 111°C in N_2 atmosphere. With the liberation of CO at 255°C a mixture of $CuCO_3$ and Na_2CO_3 is reported along with conversion of carbonates to oxides at 349°C with CO_2 liberation. K_3 $Fe^{III}(C_2O_4)_3.3H_2O$ a synthetic minguzzite, its thermal properties has been reported. At 160°C a molecule of water is immediately lost [60]. Fe_2O_3 and $K_2C_2O_4$ are formed above 160°C and up to the mentioned temperature anhydrous $K_3[Fe (C_2O_4)_3]$ so generated is reported to be stable. A stable residue is reported at 380°C when potassium oxalate decomposes at a temperature above 330°C. Formation of $KFeO_2$ has been reported at 580°C [60]. At about 500°C is mixture of K_2CO_3 and elemental Fe reported to be obtained in N_2 atmosphere. The degradation process under this atmosphere were advanced, most probable is initial reduction of Fe (III) to Fe (II):

$$2K_3 [Fe (C_2O_4)_3] \rightarrow 2K_2 [Fe (C_2O_4)_2] + K_2C_2O_4 + 2CO_2 \qquad (24)$$

Successive decomposition of complex of Fe (II) leads to formation of $Fe_3O_4/K_2C_2O_4$. CO obtained in decomposition of $K_2C_2O_4$ leads to reduction of oxide [60].

$$K_2C_2O_4 \rightarrow K_2CO_3 + CO \qquad (25)$$

Al (III) complex, $K_3 [Al (C_2O_4)_3] \cdot 3H_2O$, thermal property has been studied and reported [69, 70]. The thermal decomposition of the corresponding ammonium complex, $(NH_4)_3[Fe^{III}(C_2O_4)_3] \cdot 3H_2O$, has been reported by TG/DTA in air and N_2 atmosphere [60, 71]. Along with liberation of three water molecules first step of decomposition starts at about 100 °C in air followed with ignition at 260 °C sample burning giving fine Fe_2O_3 in the second step [71]. A mixture of Fe, Fe_3O_4, and FeO has been reported as final products in N_2 atmosphere [60].

Spectroscopic and Magnetic Properties

All the three hydrates of Mn (II) oxalates are reported to have been studied using IR and Raman spectra [79]. All these three complexes have been reported to have comparable spectra within 2000 and 400 cm^{-1} only with minor differences in the intensities. In the higher energy region of the spectra differences are observed where O-H stretching modes are present. Spectral difference in the high energy region helps in differentiating between them [79].

Identical IR and Raman patterns are reported for two of the dihydrated modifications of Fe (II) oxalate [46]. A close similarity has been reported between Raman spectrum of α-Fe $(C_2O_4) \cdot 2H_2O$ with natural humboldtine [72, 73]. Both IR and Raman for Fe (III) oxalate, was investigated [46]. Analyzing both the spectra of α-Fe $(C_2O_4) \cdot 2H_2O$ and α-Mn $(C_2O_4) \cdot 2H_2O$, the differences in the H-bond structure of both were reported.

A direct demonstration of Jahn-Teller effect has been reported on the cation Fe (II) along with the spectral properties of ν(Fe–O) stretching band[80]. It has been reported from the comparison of structures the spectra of Cu $(C_2O_4) \cdot 0.2H_2O$ is same to that of other mentioned complexes [81]. The zeolitic water in lattice structure gives a wide and vague IR-absorption at 3575 cm^{-1} without any Raman counterpart at high frequency region [81].

For synthetic $Na_2Cu (C_2O_4)_2 \cdot 2H_2O$ both the IR and Raman were reported without considering structural database [82]. It was reported to be reinvestigated comparing the related salts of potassium and ammonium. Fujita et al. reports IR spectra of synthetic $K_3 [Fe (C_2O_4)_3] \cdot 3H_2O$ [83].

Electronic Spectra

It has been reported that absorption spectra of metal oxalates were done in water and sometimes due to presence of excess amount of oxalate stabilizes the complex [4]. Few complexes of oxalato have been reported to be studied using reflectance spectroscopy or in pure crystalline form or being doped in a matrix. The complex of Na Mg $[Al (C_2O_4)_3] \cdot 9H_2O$, where Al(III) gets replaced by Fe (III) is reported to absorb polarized light allowing the study of electronic transitions for the cation. The Dq parameter (1500 cm^{-1}) found to be same to that of aqua complex [44]. The UV–VIS–NIR reported for $K_3 [Fe (C_2O_4)_3]$ shows two bands of absorption at 644 and 924 nm due to ligand-to-metal charge transfers [84]. For $Cu(C_2O_4) \cdot 0.2H_2O$, a strong and wide band has been reported in visible range reported at 760 nm and another strong one at 300 nm [81].

Magnetic Properties

Oxalate-metal groups linear chains are reported to be the striking characteristic of the M^{II} = 3d transition elements dihydrate oxalate systems with strong antiferromagnetic coupling. It has been reported that the neighboring groups are attached with weak H-bonds. A 1D – quasi magnetic property has been reported for these systems with important correlation along b axis above the transition temperature to 3D magnetic ordering [85, 86]. For synthetic oxalato complexes magnetic studies has been reported [4]. The magnetic susceptibility for α-Mn $(C_2O_4) \cdot 2H_2O$ and Mn $(C_2O_4) \cdot 3H_2O$ has been reported to be in range of 25–300 °C During the study of the thermal decomposition of α-Mn $(C_2O_4) \cdot 2H_2O$ and Mn $(C_2O_4) \cdot 3H_2O$, and in order to follow the magnetic behavior of the degradation products, magnetic susceptibility measurements in the temperature range 25–300 °C when thermal degradation takes place. Valuable magnetic moment has been reported to be 5.92 Bohr magnetons with Weiss temperature reported to be θ = −13 K. Antiferromagnetic property has been reported to be confirmed if Weiss constant has a negative value [65]. At a temperature between 1.5 and 295K for α-Mn $(C_2O_4) \cdot 2D_2O$ magnetic structure has been reported using neutron diffraction and through magnetic susceptibility measurements(4.2–295 K).The complex reports a distinctive Curie-Weiss property with S = 5/2 and θ = −11.2 K above 80K. It has been reported the Neel temperature T_N around 2.6 K and around 10K intra-chain magnetic interaction begins [86]. For [Mn $(C_2O_4) (H_2O)_2] \cdot H_2O$, Curie–Weiss reported above 30 K with θ = −20.36 K. By specific heat measurements Néel temperature of 2.82 K has been reported. Around 12K long range ordering below T_N has been reported to develop also 1-D intra-chain magnetic interactions takes place [16]. Magnetic structure was reported to be studied for α-Fe $(C_2O_4) \cdot 2D_2O$, at temperature 4.2 and 300 K via neutron diffraction [87] earlier magnetic susceptibility (1.3–300 K) for the complex has been reported [88]. Above 80K complex has been reported to show Curie–Weiss law with θ = −25.4 K [88], and Néel temperature at 16 K [87]. At a temperature range of 1.4 to 293 K [89] using neutron diffraction and EPR measurements [90] magnetic property of α-Co $(C_2O_4)\cdot 2H_2O$ has been reported. Using heat capacity studies Neel temperature was reported to be at 6.2 K [90]. The magnetic property of β-Co (C_2O_4) $\cdot 2H_2O$ has been reported [91]. A typical behavior has been reported above 100 K for the complex with θ-values between −31.8 and −40.7 K, for samples and Néel temperature has been reported at 7.5 K. Magnetic behavior so reported is found to be constant with the antiferromagnetic ordering and the hysteresis measured as function of magnetic field reports that the intra and inter connections. Although the observed general magnetic behavior is totally consistent with the existence of antiferromagnetic ordering, the hysteresis measured in magnetization experiments as a function of the magnetic field reveals that the intra- and inter-chain magnetic interactions tilt the spins, hence leading to distortion of antiferromagnetic order reporting a weak ferromagnetic property [91]. [Co (C_2O_4) (H_2O) $_2]\cdot 2H_2O$, has been reported to follow Curie–Weiss law above 70 K, having θ = −68.8 K showing a maxima at 26K for curve of magnetic susceptibility vs. temperature curve [18]. For both the modifications of $NiC_2O_4\cdot 2H_2O$, the magnetic property has been reported. Nanorods of α-$NiC_2O_4\cdot 2H_2O$, has been reported to fit in Curie–Weiss law above 100 K having θ of −83.4 or −90.4 K, which reports to depend on process of preparation, and the susceptibility at 40 K [92]. It has been reported that its β modification follows Curie–Weiss law at 100 K and above with θ between −76.9 and −99.8 K, indicating importance of antiferromagnetic exchange interactions. The temperature range between 3.3 and 43 K, reports weak ferromagnetic properties [93]. For the cis-$NiC_2O_4\cdot 2H_2O$ form, observation of the magnetic properties have been reported in temperature range of 2.5–255 K and compared with magnetic properties of the α-$NiC_2O_4\cdot 2H_2O$ [19]. The magnetic property of Cu (C_2O_4) $\cdot nH_2O$ has also been reported, which is an example of strong antiferromagnetic linear chain coupled spin doublets [13, 14, 85, 94, 95]. EPR studies as a function of temperature and frequency for the complex have been reported [96]. For Na_2 [Cu $(C_2O_4)_2$] $\cdot 2H_2O$, weak antiferromagnetic coupling has been reported, from the interactions in the chains of stacked [Cu $(C_2O_4)_2]^{2-}$units, as shown in Figure 6(A) [36]. The complexes K_2 [Cu $(C_2O_4)_2$] $\cdot 2H_2O$ and $(NH_4)_2$[Cu $(C_2O_4)_2$] $\cdot 2H_2O$

stoichiometrically related reports Curie–Weiss behavior having a small θ-values of 0.7 K for potassium and 0.6 K for ammonium salt, and EPR reports magnetic interactions among metallic centers [97], but the coupling reported to be weaker than that of the sodium salt. Figure 6(B), shows interactions between K⁻ and NH₄⁻ in [Cu $(C_2O_4)_2$]²⁻ and [Cu $(C_2O_4)_2(H_2O)_2$]²⁻ units placed in a way that interactions occurs via weak oxalate bridges at a distance more than 5 Å, but in Na salt, the arrangement helps in δ-type overlapping of the magnetic orbitals [36]. For synthetic K_3 [Fe $(C_2O_4)_3$]·3H₂O, magnetic susceptibility reported to be between 80 and 300 K, with Curie–Weiss behavior with θ = −3.0 K, reporting weak magnetic interactions among Fe (III) centers [98]. Earlier magnetic studies for the complex reports anisotropy of the $6A_{1g}$ ground state obtained using magnetic susceptibility measurements for the first time [99].

EPR spectra of this complex have been reported at both X- and Q-band frequencies in the temperature range between 77 and 300 K [100]. It's related complex, $K_3[Fe(C_2O_4)_3]$, the X-band ESR spectrum at room temperature reports to show same properties [84] as that for its trihydrated equivalent [100]. Intensity of magnetization is reported to vary linearly with the field applied backing up the paramagnetic nature of complex at room temperature [84].

TRANSITION METAL OXALATES AS POTENTIAL FUTURISTIC MATERIALS FOR EFFICIENT ENERGY STORAGE

There has been an unquestionable need for development of clean, sustainable and highly performing energy related storing and converting technologies owing to rapid depletion of fossil fuels and exponential energy demand [101-103]. This serious demand has immensely fuelled research in developing and improving various energy storing and converting devices namely batteries [104-106], supercapacitors [107-109], flow batteries [110-112] etc. Lithium ion batteries (LIBs) and supercapacitors (SCs), are energy storage devices, with broad applications in electronic vehicles and portable electronic devices due to their outstanding energy and power densities and high cycle stabilities [113]. The electrochemical performance of an electrochemical system highly relies on use of suitable electrode materials for their performance. Therefore, several efforts have been made by different researchers for performance analysis of different synthesized material along with further modifications of their morphology, chemical structures and composite material compatibility to enhance the overall performance of the system.

Amongst the variety of materials investigated transition-metal oxalates for energy applications have received extensive research interest recently owing their tunable properties. Apart from possessing high surface area and excellent physical properties transition metal oxalates can be tuned to desirable structure for enhanced activity. Transition-metal oxalates are compounds having to the transition-metal ion coordinated to the oxygen atoms of oxalate. The co-ordination of oxalate ions to transition metal ions are can be achieved through CO_2 thus establishing them as carbon sinks which can further contribute to a sustainable green future [114]. Different cost effective, sustainable and scalable synthesis methods of different morphologies for enhanced electrochemical properties have already been reported by various groups [114-116].

Furthermore, multimetal oxalates are found to reflect enhanced redox reactivity and superior conductivity compared to their unary counterparts for electrochemical applications [117-118]. Transition metals are often coupled with carbon materials to improve their performance found to degrade due to large volume expansion owing to electrochemical processes [119]. Transition metal oxalates are also considered feasible precursor sources for synthesis of their respective oxides reflecting efficient electrochemical behavior though thermal degradation [120-121]. Thus being established as versatile potential materials for energy storage and conversion, there has been significant increase in the amount of research and publications to explore all possibilities for the same (Figure 8).

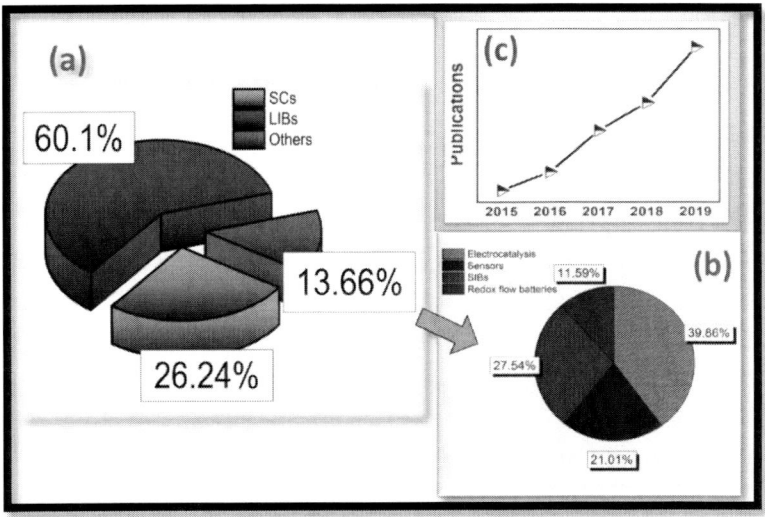

Figure 8. Research and publications on Oxalate materials for energy storage and conversion.

This section of the chapter will give an in- depth and focused insight to the electrochemical performance of the TMOs and multimetal oxalates. The use of different oxalates and multimetal oxalates in LIBs, SCs and RFBs will be summarized and compared. Finally, a conclusive remark for use of TMOs and multimetal oxalates as potential futuristic materials for efficient energy storage will be elucidated for a better understanding of the readers.

Li Ion Batteries

Li ion batteries have proven to be the most encouraging candidates for efficient electrochemical energy storage owing to its high specific energies in term of power and energy assisted with lower reduction potentials. The fast ion mobility of Li ions offers the platform for enhanced electrochemical activity. Li ion batteries have been successfully commercialized in a wide range of applications starting form mobile phones to electric vehicles paving the way for a green sustainable future.

However, the limitation to their specific capacities demand modifications in conventionally used electrode materials and electrolyte to achieve higher energy densities for next generation technologies [122-123].

Oxalates have proved to be promising materials for effective Lithium storage ability both as positive and negative electrodes with excellent retention. The ease of synthesis and tunable morphology offers scope of enhanced electrochemical activity thus establishing them as favorable candidates for Li storage. Many TMOs and TMMOs based on different transition metals have been thoroughly studied for Li storage and there has been significant progress and reports of the same [124-130].

Recently Yao et al. [131] reported a very novel fluoro-oxalate cathode material for $Na_2Fe-(C_2O_4)$ F_2 with almost 100% theoretical capacity at an average voltage of 3.6 V with an excellent 0.67Li^+ reversibility up to 50 cycles which might be improved with varying potential window establishing oxalates as efficient polyoxyanion positive electrode material. Extending the focus on novel electrode material Yao et al. and group very recently reported an oxalate cathode for lithium ion batteries with combined cationic and polyanionic redox. The cells demonstrated an excellent 1.25 Li insertion under different current densities (50 - 500 mA g^{-1}) in the potential window of 2.0 -4.5 V. The high Li ion exchange was attributed to co-existence of cationic and anionic redox which not only enhanced the capacity, but also accomplishes excellent electrochemical durability. This dual redox strategy by oxalates could pave the way for different potential research for next generation high capacity batteries [132].

Zhang et al. further studied multilayer Fe oxalate [133] stable channels and Fe oxalates/graphene sheet [134] composite for anode materials of Li ion batteries. While the multilayer FeC_2O_4 displayed a high discharge capacity of ~1521.2 mAh g^{-1} for the first cycle, with a satisfactory capacity retention of 65.30% for the 200th cycle at 1C the FeC_2O_4/graphene sheets composites reflected a reversible capacity of 916.49mAh g^{-1} and capacity retention of 66.9% at 1C after 50 cycle respectively. Zhang et al. further studied different single and mixed phases of Fe oxalate for stable Li^+ diffusion. The study reflected the mixed phase α@βFeC_2O_4 provided short diffusion paths for Li$^+$diffusion with a very high reversible capacity of 1073.17 mA g^{-1} at 0.5 V after 100 cycles Thus it was inferred that crystal structure can certainly enhance the electrochemical properties of oxalates [135].A report by the similar group explored bimetallic Fe-Cu oxalates with multiwall carbon nanotubes for anode materials in LiBs. The bimetallic Fe$_{2/3}$Cu$_{1/3}$C2O4.xH$_2$O/MWCNTs displays a good lithium storage performance with stable capacity of 530mAh^{-1}after 200 cycles at 0.5 V [136]. The series of studies by Zhang et al. clearly provides a clear illustration of various modifications that can be adopted both in terms of crystal geometry and morphology along with metal and carbon coupling to enhance the performance of electrode materials.

Park et al. proposed a newly synthesized SnC$_2$O$_4$as anodes in LIBs. A significant improvement in the electrochemical properties was seen after dispersion of rGO. The highly achieved conductivity achieved by rGO coupling permitted high reversible lithiation-delithiation processes achieving a very enhanced charge capacity of 467 mAh g^{-1} at 10 C-rate (10 A g^{-1}) [137]. Xu et al. studied cobalt oxalate CoC$_2$O$_4$nanorods for anode material for LIBs. The one dimensional porous nanostructure was found to exhibit a high reversible specific capacity besides excellent cycling stability (924 mA h g^{-1} at 50 mA g^{-1}after 100 cycles and 709 mA h g^{-1}at 200 mA g^{-1}after 220 cycles) [138]. Further studies on CoC$_2$O$_4$by Wu et al. focussed on the role of Mn in multi metal MnCoC$_2$O$_4$ system. With optimum ratio of Mn/Co urchin like Mn$_{0.33}$Co$_{0.67}$C$_2$O$_4$system which delivers a high discharge capacity of 734 mAh g^{-1}at 1 A g^{-1} and 414 mAh g^{-1} at 5A g^{-1}, respectively [139]. Cylindrical and rod-like nanostructures of CuC$_2$O$_4$.xH$_2$O was synthesized by Kanget al.using different methods. It was observed that after 100 cycles at 200 mA g^{-1}, the cylindrical CuC$_2$O$_4$.xH$_2$O reflected a discharge capacity of 920.3 mA h g^{-1}and the retention of 970.0 mA h g^{-1} which was inferior in terms of discharge capacity but superior as of higher retention than those of the rod-like aggregates of CuC$_2$O$_4$.xH$_2$O with initial discharge capacity of 1211.3 mA h g^{-1}and the retention of 849.3 mA h g^{-1} [140]. In a separate studyNiC$_2$O$_4$.2H$_2$O/rGO was found to show a higher capacity of 933 mA h g^{-1}, efficiency 87.5% compared to bare NiC$_2$O$_4$.2H$_2$O electrode with discharge capacity 805mA h g^{-1}, efficiency 72.6% [141]. The research outcomes and efficiency of metal oxalates for LIBs reported in recent years are presented in a tabular form in Table1.

Supercapacitors

Supercapacitors (SCs), also electrochemical capacitors are seen to have tremendous application potential attracting significant research as sustainable energy storing devices. SCs deliver high power density and rapid charge discharge [142-143]. Their properties are established to lie between conventional capacitors and batteries. SCs basically store charge ions adsorption on the surface of the electrode or via redox reaction near the surface namely pseudo capacitance. The performance of SCs are greatly dependent on use of suitable electrode materials [144-145]. The search for a cost effective and high performing electrode material has always been challenge which brings in picture oxalate materials for the same. Compared to the commonly used transition metal oxide, TMOs are cheap and their large surface area and tunable geometry further strengthens their stand for use in SCs.

Zhao et al. *very* recently reported two dimensional porous nickel oxalate nanosheets structures in aqueous electrolytes for high performing electrodes in supercapacitors. The high porosity and active surface area helped in achieving a tremendous capacity of 1134.024 C·g^{-1}at 1.0Ag^{-1}with a capacitive retention of 94.3% even after 5000 cycles in a high current density of 10 Ag^{-1}. When assembled into an asymmetric device it delivered a high energy density 35.7 Wh kg^{-1} at 900 W kg^{-1} [146]. Similar kind of work was also carried out by Pu et al. on copper oxalate (CuC_2O_4). The 2D interconnected nanosheets reflected a capacity of 1.631 Fcm^{-2} at the current density of 1.2 mA cm^{-2}along with excellent capacitive retention of 98.5% of its original capacity after 2000 cycles at 6.0 mAcm^{-2}. When fabricated to a device the energy density reached as high as energy density of 17.675 Wh kg^{-1}at 900 W kg^{-1} [147]. Chenget al. reported a one-step anodization of CuC_2O_4on copper foil which reportedly achieved a magnificent capacitive retention of 92% even after 100000 cycles while delivering a capacity of 1269 F g^{-1}at the current density of 6 A g^{-1} [148].

Liu et al. synthesized olive like MnC$_2$O$_4$on graphene sheets using ascorbic acid as a reducing agent and reported initial foundations of electrochemical properties of MnC$_2$O$_4$. MnC$_2$O$_4$uniformly dispersed on graphene sheets enhanced the electrochemical behavior of bare MnC$_2$O$_4$achieving a capacitance of 122 Fg^{-1} which was twice compared to bare MnC$_2$O$_4$ capacity of 53Fg^{-1}. The capacitive retention of 94.3% after 1000 cycles at 0.5 Ag^{-1} was achieved owing to the fast faradic reactions promoted by high conductivity of graphene sheets [149]. The initial foundation led by Liu and group subsequently led to expansion in research interest for Mn based oxalates and very recently a bimetallic Mn oxalate from low grade Mn ore was reported by Pattnaik et al. Leaching parameters were optimized and rod shape crystalline Fe$_2$Mn (C$_2$O$_4$)$_2$(OH)$_3$(H$_2$O)$_3$]. H$_2$O. The optimal Fe:Mn ratio was found to be 0.67 which exhibited specific capacitance of 86.9 Fg^{-1} and energy density of15.44WhKg^{-1} [150].Multimetal cobalt-manganese-nickel oxalates Co$_{0.5}$Mn$_{0.4}$Ni$_{0.1}$C$_2$O.nH2Omicropolyhedrons were also reported by Zhang et al. by a simple room temperature co-precipitation method. A very high specific capacitance of achieved 990 Fg^{-1} was acheived at 0.6 Ag^{-1} and 600 Fg^{-1} at current density as high as 4.0 Ag^{-1}. An excellent cycling stability electrode exhibited stable cycling performance with a specific even after 6000 cycles was achieved. Further assembling into solid state flexible supercapacitor a very superior energy density of 0.46 mWh cm^{-3} at 0.5 mAcm^{-2} was achieved reflecting the suitability of the material [151]. In a report by Heet al. tremella-like NiC$_2$O$_4$@NiO core/shell hierarchical nanostructures were successfully synthesized on Nickel foam by hydrothermal method followed by electrochemical activation for binder free electrode in supercapacitor devices. The core shell morphology helped in producing shorter diffusion pathways and better conductivity for enhanced performance. The morphology assisted binder free NiC$_2$O$_4$@NiO structure produced a very high capacitance of 2287.09 Fg^{-1} at 1 Ag^{-1}with a capacity retention of 95% after 10000 cycles [152]. All the above studies highlight the importance of morphology along with conductivity and large surface area of oxalate materials can create shorter diffusion paths for better electrochemical performance. The performance of the above mentioned oxalate materials have been compared in a tabular form in Table 2.

Redox Flow Batteries

Redox flow batteries are another class of energy storing devices wherein the power output and the energy output can be obtained separately as per end user applications. The flowing electrolyte volume and the exposure to the stacks can be manipulated to get desired output either in form of energy or power [153]. Oxalate materials have been used differently as per need in redox flow batteries. Incorporation of oxalate anions with metal cations form metal oxalates thereby enhancing the electrochemical performance. Oxalates are used as supporters to redox active species in flow batteries. They are used as chelating agents, stabilisers and electrolytes for flow batteries.

Oxalate was identified as a chelating agent in Br_2/Br flow batteries. The chelate with Fe^{3+}/Fe^{2+}-oxalate was inferred to be unsuitable due to constrained oxalate formation owing to low solubility of sodium oxalate. Another reason of concern was low cyclic stability that was attributed to the degradation of the ferrous form [154]. Vanadium redox batteries use oxalic based additives as stabilisers [155,156]. The basic purpose of the stabilisers render stabilisation of the vanadium supersaturated solution and restrain precipitation. However, the oxidation of the used oxalates to produce CO_2 hinder the electrolyte flow questioning the long term stability of the same [157]. As electrolyte, oxalic acid has been used as effective cathodic electrolyte. The capacities of the oxalic acid based electrolytes are found to be superior to sulphuric acid based systems. The ability to store higher concentrations of V^{4+} ions helps augmenting the electrochemical activity with the help of more participating vanadium ions. Lee et al. [158] reported capacities of 21.6/17.6 Ah L^{-1} at a current density of 30 mA cm^2

Table 1. The research outcomes and efficiency of metal oxalates for LIBs reported in recent years

Year/Authors	Material	Synthesis method	Morphology	Capacity	Cyclic stability
[131] Yao et al., 2020	Na_2Fe-$(C_2O_4)F_2$	Hydrothermal	-	70 mAh g^{-1} at an average discharge potential of 3.0 V	0.67 Li$^+$/lithium recovery in 50 cycles
[132] Yao et al., 2020	$Li_2Fe(C_2O_4)_2$	Hydrothermal	polyhedral	150 mAh g^{-1}	-
[134] Zhang et al., 2019	FeC_2O_4/Gs composite	Solvothermal process	$Fe_2C_2O_4$/Gs= ribbon like particles stacked in layers $Fe_2C_2O_4$/GS= rod like particles	α-$Fe_2C_2O_4$/Gs= Specific cap- 916.49mAh at 1C β- $Fe_2C_2O_4$/Gs= Specific cap- 550mAh/g after 200 cycles	α-$Fe_2C_2O_4$/Gs=66.9% retention at 1C after 50 cycles
[133] Zhang et al., 2018	FeC_2O_4	Liquid phase precipitation method	Rod like structure	1521.2 mAh/g at 1C	63.29% at 1st cycle and 65.30% for 200 cycle at 1C
[135] Keyu Zhang et al., 2019	FeC_2O_4	Solvothermal method	α@β- FeC_2O_4= homogeneous pores on the surface of the pristine particles	1073.17mAh/g at 0.5A/g and 841.24 mAh/g at 5A/g	79.86% retention after 100 cycles at 0.5A/g 69.02% retention after 100 cycles at 5A/g
[136] Zhang et al., 2020	$Fe_{2/3}Cu_{1/3}C_2O_4.xH_2O$/MWCNTs	Solvothermal and self-assembly technique	$CuC_2O_4.yH_2O$ = spherical shape $FeC_2O_4.2H_2O$ = microrod Both are uniformly dispersed	530mAh/g at 0.5A/g and 482mAh/g at 2A/g	530mAh/g at 0.5A/g after 200 cycles

Table 1. (Continued)

Year/Authors	Material	Synthesis method	Morphology	Capacity	Cyclic stability
[137] Park et al., 2017	SnC_2O_4/rGO composite	Hydrothermal with SA-LBL process	Plate like particles	116mAh/g at 100mA/g	620mAh/g after 200 cycles at 100mA/g
[138] Xu et al., 2015	CoC_2O_4	Facile water controlled coprecipitation method	Nanorods	924 mAh/g at 50 mA/g	924 mAh/g at 50 mA/g after 100 cycles and 709 mAh/g at 200mA/g after 200 cycles
[139] Wu et al., 2015	$Mn_{0.33}Co_{0.67}C_2O_4$	Template free chemical co-precipitation method	Urchin like structure	924 mAh/g at 500mA/g	83% retention over 300 cycles at 500mA/g
[140] Kang et al., 2013	$CuC_2O_4 \cdot xH_2O$	Hydrothermal and solvothermal method	Cylinder and rod like nanostructure	Cylinder like = 970/100th cycle at 200mA/g and 809.5/100th cycle at 500mA/g Rod like = 849.3/100th cycle at 200mA/g and 687.8/100th cycle at 500mA/g	-
[141] Oh et al., 2016	NiC_2O_4/rGO composite	Hydrothermal with SA-LBL process	Nanorods	586mAh/g at 10C	87.5% for 1st cycle and 85% retention for 100 cycles at 10C

Table 2. The performance of oxalate materials as supercapacitors

Year/Authors	Material	Synthesis methods	Morphology	Specific cap	Cyclic stability	Device performance
[146] Zhao et al., 2020	2D porous Nickel Oxalate	Hydrothermal Method	Thin sheets	1134.024 C/g at CD = 1.0 A/g	94.3% after 5000 cycles at 1 A/g	ASC device: Ni-OA thin sheets//active carbon Energy density = 35.7 Wh/kg Power density = 900 W/kg 92.5% capacitance retention after 10000 cycles.
[147] Pu et al., 2018	Cobalt Oxalate	Hydrothermal Method	Layered structure with abundant pores and thin sheets	High area specific capacitance = 1.631 $F.cm^{-2}$ at CD = 1.2 $mA\ cm^{-2}$	80.6% capacitance retention at 1.2 $mA\ cm^{-2}$	ASC device: Energy density = 17.675 Wh/kg Power density = 900 W/kg
[148] Cheng et al., 2016	Cobalt oxalate anchored Co foil electrode	Facile one step anodization	Nanorods	1269 F/g at CD = 6A/g	91.9% retention after 100,000 cycles at 6A/g	ASC device: Specific capacitance = 77F/g at CD = 0.5A/g Energy density = 27.3Wh/kg Power density = 0.429 kW/kg
[149] Liu et al., 2014	MnC_2O_4/Graphene composite	Facile hydrothermal method	Olive-like structures	122 F/g at CD = 0.5A/g	94.3% retention after 1000 cycles at 0.5A/g	-
[150] Pattnaik et al., 2019	Fe-Mn bimetallic oxalate	Leaching of low grade Mn ore	Rod shape	86.9 F/g at CD = 0.1 A/g	94.3% after 1000 cycles at 0.1 A/g	
[151] Zhang et al.,2015	$Co_{0.5}Mn_{0.5}$ $Ni_{0.1}C_2O_4.nH_2O$	Chemical co-precipitation method	Uniform micropolyhedron	990 F/g at CD = 0.6A/g		SASC device: Specific capacitance = 86.3$mF.cm^{-2}$ 98.6% retention 6000 cycles Energy density = 0.46$mWh.cm^{-3}$
[152] He et al.,2017	NiC_2O_4@NiO	Facile hydrothermal method followed by electrochemical activation process	Tremella like morphology	2287 F/g at CD = 1A/g	95% retention after 10000 cycles at 1 A/g	-

cycling within a potential window 0.8 and 1.6 V. Though some reports of oxalates in use of redox flow batteries are highlighted extensive research is needed in the direction for their possible use in redox flow batteries for electrochemical properties enhancement and long term stability of the system.

CONCLUSION AND OUTLOOK FOR RATIONALE USE OF OXALATE IN ENERGY MATERIALS

Transition metal oxalates have mainly found their applications in Lithium ion batteries. The performance of the oxalates in LIBs is quite alluring but faces the constraint of low reversibility and stability. Beyond LIBs the use of the Oxalates has been promising as electrode materials for supercapacitors in aqueous electrolytes. The enhancement of the electrochemical properties can be facilitated by optimization of the particle symmetry including dimension reduction, composite development, morphology control, doping, binder and electrolyte selection. TMOs in redox flow batteries acts in an auxiliary role to enhance the active redox species.

The enhancement of the electrochemical properties can be facilitated by optimization of the particle symmetry including dimension reduction, composite development, morphology control, doping, binder and electrolyte selection. TMOs in redox flow batteries acts in an auxiliary role to enhance the active redox species. In short, TMOs maturity as energy storage materials still stands new and their use in energy devices demand more research understanding how their behavior along with strategies to improve their electrochemical performance. Nevertheless, the several reports of oxalate materials show promising results and with a better understanding and focused research TMOs possess extremely high potential as potential futuristic materials for efficient energy storage.

REFERENCES

[1] Strunz, H., Nickel, E.H. (2001). *Strunz mineralogical tables.* 9th ed. Stuttgart: Schweizerbart'scheVerlagsbuchhandlung.
[2] Piro, O.E., Baran, E.J. (2018). Crystal chemistry of organic minerals – salts of organic acids: the synthetic approach, *Crystallography Reviews.*
[3] https://en.wikipedia.org/wiki/Oxalate.
[4] Krishnamurthy, K.Y., Harris, G.M. (1961). *Chem. Rev,* 61: 213.
[5] Baran, E.J., Monje, P.V. In *Biomineralization. From Nature to Application, Metal Ions in Life Sciences,* A. Sigel, H. Sigel, R.K.O. Sigel (Eds.), Vol. 4, pp. 219–254, Wiley, Chichester (2008).
[6] Deyrieux, R., Berro, C., Péneloux, A. (1973). *Bull. Soc. Chim. Fr.,* 25.
[7] Lagier, J.-P., Pezerat, H., Dubernat, J. (1969). *Rev. Chim. Minér.,* 6: 1081.
[8] Echigo, T., Kimata, M. (2010). *Can. Mineral.,* 48: 1329.
[9] Walter Levy, L., Perrotey, J. (1970). *Bull. Soc. Chim. Fr.,* 1697.
[10] Walter Levy, L., Perrotey, J., Visser, J.W. (1971). *Bull. Soc. Chim. Fr.,* 757.
[11] Huizing, A., van Hal, H.A.M., Kwestroo, W., Langereis, C., Loosdregt, P. Van. (1977). *Mater. Res. Bull.,* 12: 605.
[12] Figgis, B.N., Martin, D.J. (1966). *Inorg. Chem.,* 5: 100.
[13] Michalowicz, A., Girerd, J.J., Goulon, J. (1979). *Inorg. Chem.,* 18: 3094.
[14] Lethbridge, Z.A.D., Congreve, A.F., Esslemont, E., Slawin, A.M.Z., Lightfoot, P. (2003). *J. Solid State Chem.,* 172: 212.
[15] Wu, W.Y., Song, Y., Li, Y.Z., You, X.Z. (2005). *Inorg. Chem. Commun.,* 8: 732.
[16] Fu, X., Wang, C., Li, M. (2005). *Acta Crystallogr.,* E61, m1348.
[17] Garcia a-Couceiro, U., Castillo, O., Luque, A., Beobide, G., Roman, P. (2004). *Inorg. Chim.Acta,* 357:339.

[18] Paredes-García, V., Rojas, I., Venegas-Yazigi, D., Spodine, E., Resende, J.A.L.C., Vaz, M.G.F., Novak, M.A. (2011). *Polyhedron*, 30: 3171.
[19] Molinier, M., Price, D.J., Wood, P.J., Powell, A.K. (1997). *J. Chem. Soc., Dalton Trans.*, 4061.
[20] Gavris, G., Stoia, M., Stanasel, O., Hodisan, S. (2010). *Chem. Bull. "Politehnika", Univ. Timisoara*, 55: 143.
[21] Castillo, O., Luque, A., Julve, M., Lloret, F., Román, P. (2001). *Inorg. Chim. Acta*, 315: 9.
[22] Castillo,O., Luque, A., Román, P., Lloret, F., Julve, M. (2001). *Inorg. Chem.*, 40: 5526.
[23] Lu, J.Y., Schroeder, T.J., Babb, A.M., Olmstead, M. (2001). *Polyhedron*, 20: 2445.
[24] Lin, X.R., Ye, R.Z., Liu, B.S., Wei, C.X., Chen, J.X. (2006). *Acta Crystallogr.*, E62, m2130.
[25] Li, P.Z., Xu, Q. (2009). *Acta Crystallogr.*, E65, m508.
[26] Wang, J., Hou, Y., Fang, Z. (2010). *ActaCrystallogr.*, E66, m1229.
[27] Palacios, D., Wladimirsky, A., D'Antonio, M.C., González-Baró, A.C., Baran, E.J. (2011). Sp*ectrochim. Acta, Part A*, 79: 1145.
[28] Gleizes, A., Maury, F., Galy, J. (1980). *Inorg. Chem.* 19: 2074.
[29] Kolitsch, U. (2004). *Acta Crystallogr.* C60, m129.
[30] Pannhorst, W., Löhn, J. (1974). *Z. Kristallogr.* 139: 236.
[31] Palmer, W.G. (1954). *Experimental Inorganic Chemistry*, Cambridge University Press, Cambridge.
[32] Johnson, R.C. (1970). *J. Chem. Educ.*, 47: 702.
[33] H.S. Booth (Ed.). *Inorganic Synthesis*, Vol. 1, p. 36., McGraw-Hill, New York (1939).
[34] Herpin. P. (1958). *Bull. Soc. Franç. Minéral. Cristallogr.*, 81: 245.
[35] Piper, T.S., Carlin, R.L. (1961). *J. Chem. Phys.*, 35: 1809.
[36] Galwey,A.K., Mohamed, M.A. (1993). *Thermochim. Acta*, 213: 279.
[37] Hawthorne, F.C., Krivovichev, S.V., Burns, P.C. (2000). *Rev. Mineral. Geochem.*,40: 1.
[38] Peacor, D.R., Rouse, R.C., Essene, E.J., Lauf, R.J. (1999). *Can. Mineral.*,37: 1453.

[39] Rouse, R., Peacor, D.R., Essene, E.J., Coskren, T.D., Lauf, R.J. (2001). *Geochim. Cosmochim. Acta,* 65: 1101.
[40] Yuan, Y.P., Song, J.L., Mao, J.G. (2004). *Inorg. Chem. Commun.,* 7: 24.
[41] Wang, N., Yue, S.T., Liu, Y.L. (2009). *J. Coord. Chem.,* 62: 1914.
[42] Banford, C.H., Tipper, C.F.H. (1980). (Eds.). *Chemical Kinetics, Reactions in the Solid State,* Vol. 22, pp. 218–224, Elsevier, Amsterdam.
[43] Dollimore. D. (1987). *Thermochim. Acta,* 117: 331.
[44] Frost, R.L., Weier, M.L. (2004). *Thermochim. Acta,* 409: 79.
[45] Frost, R.L., Weier, M.L. (2003). *Thermochim. Acta,* 406: 221.
[46] Frost, R.L., Adebajo, M., Weier, M.L. (2004*).Spectrochim. Acta, Part A,* 60: 643.
[47] Boldyrev, V.V. (2002). *Thermochim. Acta,* 388: 63.
[48] Mohamed, M.A., Galwey, A.K., Halawy, S.A. (2005). *Thermochim. Acta,* 429: 57.
[49] Chaiyo, N., Muanghlua, R., Niemcharoen, S., Boonchom, B., Seeharaj, P., Vittayakorn, N. (2012). *J. Therm. Anal. Calorim.,* 107: 1023.
[50] Mohandes, F., Davar, F., Salavati-Niasari, M. (2010).*J. Phys. Chem. Solids,* 71: 1623.
[51] Erdey, L., Gál, S., Liptay, G. (1964). *Talanta,* 11: 913.
[52] Macklen, E.D. (1967). *J. Inorg. Nucl. Chem.,* 29: 1229.
[53] Nicholson. G.C. (1967). *J. Inorg. Nucl. Chem.,* 29: 1599.
[54] Hermanek, M., Zboril, R., Mashlan, M., Machala, L., Schneeweiss, O. (2006). *J. Mater. Chem.,* 16: 1273.
[55] Donkova, B., Mehandjiev, D. (2004). *Thermochim. Acta,* 421: 141.
[56] Donkova, B., Mehandjiev, D. (2005). *J. Mater. Sci.,* 40: 3881.
[57] Lamprecht, E., Watkins, G.M., Brown, M.E. (2006). *Thermochim. Acta,* 446: 91.
[58] Frost, R.I., Locke, A.J., Martens, W.N. (2008*). J. Therm. Anal. Calorim.,* 93: 993.
[59] Jun, L., Feng-Xing, Z., Yan-Wei, R., Yong-Qian, H., Ye-Fei, N. (2003). *Thermochim. Acta,* 406: 77.

[60] Verdonk, A.H. (1972). *Thermochim. Acta*, 4: 25.
[61] Diefallah, El-H.M., Basahel, S.N., El-Bellihi, A.A. (1996). *Thermochim. Acta*, 290: 123.
[62] Frost, R.L., Weier, M.L. (2003). *J. Raman Spectrosc.*, 34: 776.
[63] Frost, R.L. (2004). *Anal. Chim. Acta*, 517: 207.
[64] Shippey, T.A. (1980). *J. Mol. Struct.*, 67: 223.
[65] Hind, A.R., Bhargava, S.K., Bronswijk, W., Grocott, S.C., Eyer, S.L. (1998). *Appl. Spectrosc.*, 52: 683.
[66] Cadene, M., Fournel, A. (1977). *J. Mol. Struct.*, 37: 35.
[67] Clark, R.J.H., Firth, S. (2002). *Spectrochim. Acta, Part A*, 58: 1731.
[68] Petrov, I., Šoptrajanov, B. (1975). *Spectrochim. Acta, Part A*, 31: 309.
[69] Shippey, T.A. (1980). *J. Mol. Struct.*, 63: 157.
[70] Kontoyannis, C.G., Bouropoulos, N.C., Koutsoukos, P.G. (1997). *Appl. Spectrosc.*, 51: 64.
[71] D'Antonio, M.C., Mancilla, Nancy, Wladimirsky, A., Palacios, D., González-Baró, A.C., Baran, E.J. (2010). *Vib. Spectrosc.*, 53: 218.
[72] Mancilla, N., Caliva, V., D'Antonio, M.C., González-Baró, A.C., Baran, E.J. (2009). *J. Raman Spectrosc.*, 40: 915.
[73] Echigo, T., Kimata, M. (2008). *Phys. Chem. Miner.*, 35: 467.
[74] Sledzinska, I., Murasik, A., Fischer, P. (1987). *J. Phys. C: Solid State Phys.*, 20: 2247.
[75] Śledzińska, I., Murasik, A., Piotrowski, M. (1986). *Physica B*, 138: 315.
[76] De, S., Barros, S., Friedberg, S.A. (1966). *Phys. Rev.*, 141: 637.
[77] Lukin, J.A., Simizu, S., VanderVen, N.S., Friedberg, S.A. (1995). *J. Magn. Magn. Mater.*, 140–144: 1669.
[78] Romero, E., Mendoza, M.E., Escudero, R. (2011). *Phys. Status Solidi B*, 248: 1519.
[79] Vaidya, S., Rastogi, P., Agarwal, S., Gupta, S.K., Ahmad, T., Antonelli, A.M., Ramanujachary, K.V., Lofland, S.E., Ganguli, A.K. (2008). *J. Phys. Chem. C*, 112: 12610.
[80] Romero-Tela, E., Mendoza, M.E., Escudero, R. (2012). *J. Phys. Condens. Matter*, 24: 196003.

[81] Jeter, D.Y., Hatfield, W.F. (1972). *Inorg. Chim. Acta,* 6: 523.
[82] Delgado, G., Mora, A.J., Sagredo, V. (2002). *Physica B,* 320: 410.
[83] Khomyakov, AP. (1996). *ZapiskiVseross Mineral Obshch,* 125:126–132. (in Russian).
[84] Jambor, J.L., Pertsev, N.N., Roberts, A.C. (1997). *Mineral.* 82:430–433.
[85] Reed, D.A., Olmstead, M.M. (1981). Sodium oxalate structure refinement. *Acta Cryst. B,* 37: 938–939.
[86] Tazzoli, V., Domeneghetti, C. (1980). *Amer Mineral,* 65:327–334.
[87] Deganello, S., Piro, O.E. (1981). *Neues Jahrb Mineral Monatsh,* 2:81–88.
[88] Basso, R., Lucchetti, G., Zefiro, L., (1997). *Neues Jahrbuch fur Mineralogie Abhandlungen,* 2:84–96.
[89] Conti, C., Casati, M., Colombo, C., (2015). *SpectrochimActa,* 150A: 721–730.
[90] Rastsvetaeva, R.K., Chukanov, N.V., Nekrasov, Y. (2001). *Doklady Chem,* 381:329–331.
[91] Piro, O.E., Echeverria, G.A., Gonzalez-Baro, A.C., Baran, E.J. (2018). *Phys Chem Min,* 45:185–195.
[92] Baran, E.J. (2014). *J Coord Chem,* 67:3734–3768.
[93] Taylor, J.C., Sabine, T.M. (1972).*ActaCryst B,* 28:3340–3351.
[94] Atencio, D., Coutinho, J.M.V., Graesser, S.(2004). *Amer Mineral,* 89:1087–1091.
[95] Soleimannejad, J., Aghabozorg, H., Hooshmand, S. (2007). *Acta Cryst,* 63: 2389–2390.
[96] Fu, X., Wang, C., Li, M. (2005). *Acta Crystallogr,* 61:1348–1349.
[97] Christensen, A.N., Lebech, B., Andersen, N.H. (2014). *Dalton Transact,* 43:16754–16768.
[98] Rouse, R.C., Peacor, D.R., Dunn, P.J., (1986). *Amer Mineral,* 71:1240–1242.
[99] Gleizes, M.F., Galy, J. (1980). *Inorg Chem,* 19:2074–2078.
[100] Chukanov, N.V., Aksenov, S.M., Rastsvetaeva, R.K., (2015*). Chile. Mineral Mag,* 79:1111–1121.

[101] Klaysom, C., Cath, T., Depuydt, T., Vankelecom, I. (2013). *Chem. Soc. Rev.,* 42: 6959.
[102] Xiong, D., Li, X., Bai, Z., Lu, S. (2018). *Small.* 14: 1703419.
[103] Hannan, M.A., Hoque, M.M., Mohamed, A, Ayob, A. (2017). *Renew. Sustain. Energy Rev.* 69:771.
[104] Kang, B., Ceder, G. (2009).*Nature.* 458:190.
[105] Nitta, N., Wu, F., Lee, J.T., Yushin, G. (2015). *Mater. Today* 18: 252.
[106] Xu, Y., Li, Q., Xue, H., Pang, H., (2018). *Coord. Chem. Rev.* 376: 292.
[107] Sahoo, R.K., Das, A., Singh, S., Lee, D., Singh, S.K., Mane, R.S., Yun, J.M., Kim, K.H.(2019). *Prog Nat Sci-Mater.* 29:410.
[108] Arico, A.S., Bruce, P., Scrosati, B., Tarascon, J.-M., van Schalkwijk, W. (2005).*Nat. Mater.*4:377.
[109] Wang, G., Zhang, L., Zhang, J. (2012). *Chem. Soc. Rev.* 41:797.
[110] Roe, S., Menictas, C., Skyllas-Kazacos, M. (2016).*J. Electrochem.Soc.* 163: 5023.
[111] Park, M., Ryu, J., Wang, W., Cho, J. (2016).*Nat. Rev. Mater.* 2: 16080.
[112] Lobato, J., Oviedo, J., Cañizares, P., Rodrigo, M.A., MaríaMillán. (2020). *Carbon.* 156:287.
[113] Wang, T., Chen, S., Pang, H., Xue, H., Yu, Y. (2017). *Adv. Sci.* 4: 1600289.
[114] Yeoh, J.S., Armer, C.F., Lowe, A. (2018). Mater. *Today Energy.* 9: 198.
[115] Cheng, G., Si, C., Zhang, J., Wang, Y., Yang, W., Dong, C., Zhang, Z. (2016). *J. Power Sources.* 312: 184.
[116] Chen, L., Zhang, Q., Xu, H., Hou, X., Xuan, L., Jiang, Y., Yuan, Y. (2015*). J. Mater. Chem.* A. 3:1847.
[117] Wei, T.Y., Chen, C.H., Chien, H.C., Lu, S.Y., Hu, C.C., *Adv. Mater.* (2010). 22: 347.
[118] Chien, H.C., Cheng, W.Y., Wang, Y.H., Lu, S.Y. (2012). *Adv. Funct. Mater.* 22: 5038.

[119] Mukherjee, R., Krishnan, R., Lu, T.M., Koratkar, N. (2012). *Nano Energy.* 1: 518.
[120] Darbar, D., Reddy, M.V., Sundarrajan, S., Pattabiraman, R., Ramakrishna, Chowdari, S., B.V.R. (2016).*Mater. Res. Bull.* 73: 369.
[121] Ahmad, T., Ramanujachary, K.V., Lofland, S.E., Ganguli, A.K. (2004). *J. Mater. Chem.*14:3406.
[122] Nitta, N., Wu, F., Lee J.T., Yushin, G. (2015). *Mater. Today.* 18:264.
[123] Wu F., Yushin, G. (2017). *Energy Environ. Sci.,* 10: 435.
[124] Ang, W.A.E, Cheah, Y.L., Wong, C.L., Hng, H.H., Madhavi, S. (2015). *J. Alloy. Comp.,* 638:324.
[125] Wu, X., Guo, J., McDonald, M.J., Li, S., Xu, B., Yang, Y. (2015). *Electrochim. Acta,* 163:93.
[126] Leon, B., Vicente, C.P., Tirado, J.L. (2012). *Solid State Ion.* 225: 518.
[127] Feng, F., Kang, W., Yu, F., Zhang, H., Shen, Q. (2015). *J. Power Sources,* 282: 109.
[128] Aragon, M.J., Leon, B., Serrano, T., Perez Vicente, C., Tirado, J.L. (2011). *J. Mater. Chem.* 21:10102.
[129] Zhang, Y., Lu, Z., Guo, M., Bai, Z., Tang, B. (2016). *JOM* (J. Occup. Med.) 68: 2952.
[130] Ang, W.A., Gupta, N., Prasanth, R., Madhavi, S. (2012). *Appl. Mater. Interfaces,* 4:7011.
[131] Yao, W., Sougrati, M.T., Hoang, K., Hui, J., Lightfoot, P., Armstrong A.R. (2017). *Chem. Mat.* 29: 2167.
[132] Yao, W., Armstrong, A.R., Zhou, X., Sougrati, M.T., Kidkhunthod, P., Tunmee, S., Sun, C., Sattayaporn, S., Lightfoot, P., Ji, B., Jiang, C., Wu, N., Tang Y., Cheng H. (2019). *Nat.Comm.* 10:3483.
[133] Zhang, K., Liang, F., Wang, Y., Dai, Y., Yao, Y. (2019). *J. Alloys Compd.* 779: 91.
[134] Zhang, K., Li, Y., Wang, Y., Yuan, M., Dai, Y., Yao, Y. *Mat. Lett.* (2019). 238: 187.

[135] Zhang, K., Liang, F., Wang, Y., Zhao, J., Chena, X., Dai, Y., Yao, Y. (2020). *Chem.* 384: 123281.

[136] Zhang, K., Gao, G., Li, Y., Wang, Y., Xu, R., Yuan, M., Dai, Y., Yao, Y. *Mat. Lett.* (2020). 266:127476.

[137] Park, J.S., Jo, J.H., Yashiro, H., Kim, S.S., Kim, S.J., Sun, Y.K., Myung, S.T. *Appl. Mater. Interfaces.* (2017). 9:25941.

[138] Xu, J., He, L., Liu, H., Han, T., Wang, Y., Zhang, C., Zhang, Y. (2015). *Electrochim. Acta.* 170:85.

[139] Wu, X., Guo, J., McDonald, M.J., Li, S., Xu, B., Yang, Y. (2015). *Electrochim. Acta.* 163:93.

[140] Kang, W., Shen, Q.J. *Power Sources.* (2013). 238: 203.

[141] Oh, H.J., Jo, C.H., Yoon, C.S., Yashiro, H., Kim, S.J., Passerini, S., Sun, Y.K., Myung S.T. (2016). *NPG Asia Mater.* 8:270.

[142] Zheng, S., Li, X., Yan, B., Hu, Q., Xu, Y., Xiao, X., Xue, H., Pang, H. (2017).*Adv. Energy Mater.* 7:1602733.

[143] Zhang, Q.Z., Zhang, D., Miao, Z.C., Zhang, X.L., Chou, S.L. (2018). *Small* 14:1702883.

[144] Pu, T., Li, J., Jiang, Y., Huang, B., Wang, W., Zhao, C., Xie, L., Chen, L. *Dalt. Trans.* (2018). 47: 9241.

[145] Rao, C., Natarajan, S., Vaidhyanathan, R. (2004). *Angew. Chem. Int. Ed.* 43: 1466.

[146] Zhao, C., Jiang, Y., Liang, S., Gao, F., Xie, L., Chen, L. (2020). *Cryst Eng Comm.* 22:2953-2963.

[147] Pu, T., Li, J., Jiang, Y., Huang, B., Wang, W., Zhao, C., Xie, L., Chen, L. *Dalt. Trans.* (2018). 47: 9241.

[148] Cheng, G., Si, C., Zhang, J., Wang, Y., Yang, W., Dong, C., Zhang, Z.J. *Pow. Sources.* (2016). 312:191.

[149] Liu, T., Shao, G., Ji, M., Ma, Z.,*Ionics* (2014). 20:145.

[150] Pattnaik, S., Mukherjee P., Barikb R., Mohapatra M. *Hydrometallurgy.* (2019). 189:105139.

[151] Zhang, Y.Z., Zhao, J., Xia, J., Wang, L., Lai, W.Y., Pang, H., Huang, W. (2015). *Sci. Rep.* 5:8536.

[152] He, D., Liu G., Pang, A., Jiang, Y., Suo H., Zhao C. (2017). *Dalton Trans.* 46:1857.

[153] Park, M., Ryu, J., Wang, W., Cho, J. (2016). *Nat. Rev. Mater.* 2:16080.

[154] Wen, Y.H., Zhang, H.M., Qian, P., Zhou, H.T., Zhao, P., Yi, B.L., Yang, Y.S. (2006). *J. Electrochem. Soc.* 153:929,.

[155] Armand, M., Grugeon, S., Vezin, H., Laruelle, S., Ribiere, P., Poizot, P., Tarascon, J.M. (2009). *Nat. Mater.* 8:120.

[156] Luo, D.-m., Xu, Q., Sui, Z.-t., Chin. J. *J. Power. Sources.* (2004).28:94.

[157] Roe, S., Menictas, C., Skyllas-Kazacos, M. (2016). *J. Electrochem. Soc.* 163: 5023.

[158] Lee, J.G., Park, S.J., Cho, Y.I., Shul, Y.G. (2013). *RSC Adv.* 3:21347.

In: Oxalate
Editor: Elsa Kytönen

ISBN: 978-1-53618-303-0
© 2020 Nova Science Publishers, Inc.

Chapter 3

OXALATE BIOSYNTHESIS AND DEGRADATION IN PLANTS AND FUNGI

Enrique J. Baran[*]

Facultad de Ciencias Exactas,
Universidad Nacional de La Plata, La Plata, Argentina

ABSTRACT

A number of different possible synthetic pathways for the generation of oxalic acid in plants and fungi are presented and briefly discussed. An important number of studies have shown that ascorbic acid is the major substrate for the synthesis of oxalic acid in plants. Therefore, the so called "Wheeler-Smirnoff" mechanism, which explains the biosynthesis of ascorbic acid in the plant kingdom, is discussed in detail. The possible involvement of D-erythroascorbic acid in the generation of fungal $H_2C_2O_4$ is also analyzed. The general characteristics of the two major oxalate degradation enzymes, oxalate oxidase and oxalate decarboxylase, are thoroughly discussed. Both systems are Mn(II)-depending enzymes, associated with the "cupin" super family of proteins, and their reaction mechanisms are closely related. Finally, the role of oxalotrophic bacteria in the oxalate-degradation processes is also briefly commented.

---

[*] Corresponding Author's Email: baran@quimica.unlp.edu.ar.

Keywords: oxalic acid, synthetic pathways, ascorbic acid, Wheeler-Smirnoff-mechanism, oxalate degradation enzymes, oxalate oxidase, oxalate decarboxylase, oxalotrophic bacteria

INTRODUCTION

Crystalline oxalates are widely distributed in Nature, and have been observed in rocks, soil, and among a variety of living organisms, including plants and animals (Baran and Monje 2008). They are usually classified by mineralogists as "organic minerals" or "organic crystals" (Echigo and Kimata 2010; Strunz and Nickel 2001; Weiner and Dove 2003). Despite the apparent contradiction of these terms, it is now clear that living organisms generate most of these crystalline phases by employing some well-known biochemical and physiological mechanisms, usually named as "biologically induced" or "biologically controlled" mineralization (Baran 1995; Monje and Baran 2004; Veis 2008; Weiner and Dove 2003).

The most widely distributed biomineral of this type is calcium oxalate, which is especially common in the plant kingdom (Arnott 1982; Baran 2016; Baran and Monje 2008; Khan 1995; Monje and Baran 2004). The presence of other metallic oxalates is extremely rare in biological systems, although the presence of crystalline magnesium, manganese, copper, and ammonium oxalates in different forms of life has been reported, whereas soluble forms of sodium and potassium oxalates are widely distributed in plants and fungi (Baran and Monje 2008; He et al. 2014).

Crystalline calcium oxalates have been found in two different hydration states in plants, either as the monoclinic monohydrate (whewellite, $CaC_2O_4 \cdot H_2O$) or as the orthorhombic dihydrate (weddellite, $CaC_2O_4 \cdot 2H_2O$). A third hydrate, $CaC_2O_4 \cdot 3H_2O$ (caoxite), is also known (Deganello et al. 1981) and although it has never been found in plants, it may be an important precursor of the other two hydrates (Young and Brown 1982).

Whewellite is the most stable form from the thermodynamic point of view (Baran and Monje 2008; Monje and Baran 2004). Although the discussion of calcium oxalate functions in plants has been frequently controversial, some essential aspects become increasingly accepted (He at al. 2014; Nakata 2003).

Because higher plants incorporate calcium in excess to the cellular requirements and most of them, unlike animals, do not have well-developed excretory systems to eliminate such excess, they modulate the difference between the natural abundance of calcium and the very low intracellular levels required by tightly controlling the distribution of calcium and its compartmentation within cells. In this context, with a solubility product of 1.3×10^{-9} in water, calcium oxalate effectively sequesters calcium and renders it metabolically and osmotically inactive for plant cells. Consequently, many plants accumulate crystalline calcium oxalate in response to the presence of an excess of calcium (Baran and Monje 2008; Webb 1999).

On the other hand, a number of recent studies indicate that these calcium oxalate crystals do not form an inert, non-retrievable, deposit of calcium, but can act as a sort of reservoir to which plants can recur in cases of calcium deficiency. Apart of this primary function as a calcium regulator, the great variety of crystal shapes and sizes, as well as their localization in different plant tissues, suggests various other functions for calcium oxalates which might be evolved secondarily. Some of these functions include mechanical support, gravity perception, intracellular pH regulation and ion balance, detoxification of aluminum and heavy metals and even light gathering and scattering to optimize photosynthesis (Baran and Monje 2008; He et al. 2014; Nakata 2003).

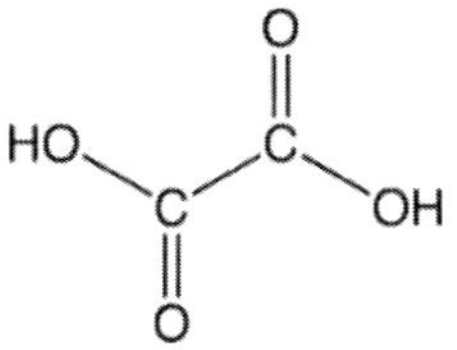

Figure 1. Schematic drawing of the structure of oxalic acid (ethanedioic acid).

Oxalic acid, ethanedioic acid (Figure 1), is the simplest of the dicarboxylic acids. The acid and some of its salts have been known since more than two hundred years. Its protonation equilibria have been often investigated; the pK values obtained at 25°C are $pK_1 = 1.27$ and $pK_2 = 4.27$ (Darken 1941), indicating that oxalic acid is a strong organic acid.

Oxalic acid is also involved in a wide range of environmental effects, including important biological and geochemical processes in soils and rocks (Adamo and Violante 2000), and has also a key importance in the carbon cycle (Baran 2016; Hofmann and Bernasconi 1998). Due to its presence in soils, together with other organic acids, it has an important effect on the bioavailability of calcium, aluminum, iron and phosphor. Together with this role in plant nutrition, oxalic acid is also involved in the weathering of soil minerals and the subsequent precipitation of insoluble metal oxalates, a process which is closely related with the survival of fungi and other species growing in the presence of potentially toxic metal compounds. It is also a toxin associated with many plant pathogens and is secreted into the plant during the infection process, participating in the degradation of the plant cells, and is then often reused by the pathogens or, finally, precipitated in the form of calcium oxalate crystals (Baran and Monje, 2008). Due to this variety of effects and also to the fact that oxalic acid is produced as a metabolic end product not only by plants and fungi but also by animals, it is not surprising that microorganisms which attack oxalate are also widely distributed in Nature.

In the first part of this review the most important aspects related to the biosynthesis of oxalic acid in plants and fungi are briefly discussed and in the second part the different forms of degradation of the acid are analyzed in a very detailed way.

BIOSYNTHESIS OF OXALATE IN PLANTS AND FUNGI

General Aspects

A number of different pathways for oxalate production have been hypothesized (Franceschi and Nakata 2005; Nakata 2003). The acid can be formed through the oxidation of glycolate and glyoxylate by the action of glycolate oxidase. This enzyme (EC 1.1.3.1) catalyzes the oxidation of glycolate to glyoxylate and that of the last species to oxalate. These two potential substrates (see Figure 2), which can be considered as effective precursors of oxalate, are formed as by products of photorespiration in photosynthetically active tissues, and the peroxisomal enzyme glycolate oxidase is fairly abundant in green tissues (Franceschi and Nakata 2005; Libert and Franceschi 1987).

Another possible route for oxalate production is again the preliminary formation of glyoxylate, by cleavage of isocitrate (Figure 2) through the enzyme isocitrate lyase (EC 4.1.3.1) (Franceschi and Nakata 2005; Libert and Franceschi 1987; Raven et al. 1982).

A third possibility may be the oxidation of oxaloacetate (Figure 2) to oxalate and acetate by the enzyme oxaloacetate hydrolase (EC 3.7.1.1) (Franceschi and Nakata, 2005; Libert and Franceschi 1987; Raven et al. 1982).

Finally, L-ascorbic acid (Figure 3) has been shown as the preferred substrate for oxalate synthesis in a wide number of plant species (Franceschi and Loewus 1995; Franceschi and Nakata 2005).

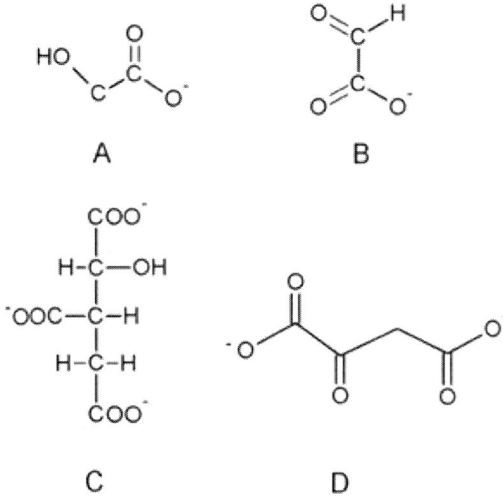

Figure 2. Some substrates proposed as possible oxalic acid precursors in plants: glycolate (A), glyoxylate (B), isocitric acid (C), oxaloacetate (D).

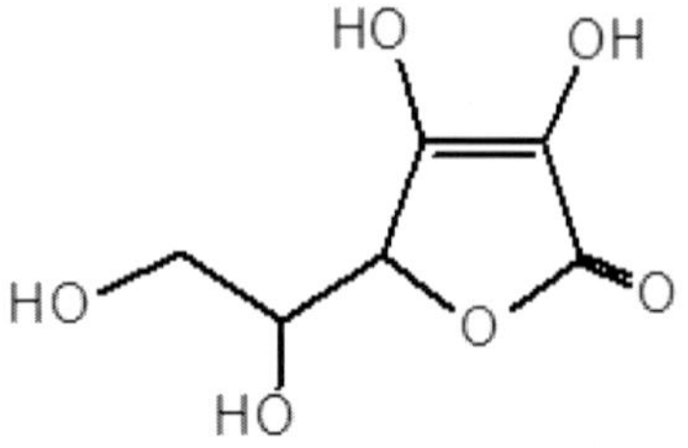

Figure 3. Schematic structure of L-ascorbic acid.

An analysis of the available information suggests that the oxaloacetate pathway is unlikely because it has only been found in a few species, and the isocitrate pathway can also be considered as minor. The glycolate pathway is a strong possibility because it is a by product of photosynthesis. However, oxalate and calcium oxalates are also formed in abundance in many non-photosynthetic plant tissues. Additionally, glycolate oxidase has not been found in some plant species generating oxalate, and other studies indicated that crystal formation is not always directly associated with photorespiration (Franceschi and Nakata 2005).

Ascorbic acid has the potential to be a good substrate for oxalic acid formation because it is present at relatively high levels and is the most abundant of the mentioned possible precursor substrates. It is also found in all tissues, green and no green (Smirnoff et al. 2001).

Ascorbic acid is found in all higher plant species. Animals which possess the capacity to oxidize L-gulono-1,4-lactone can synthesize ascorbic acid; whereas that which lack this capacity (fish, birds and mammals including primates) must find a nutritional source in order to survive (Loewus 1999).

In the case of fungi, information regarding the biosynthesis of oxalate remains fragmentary, although the participation of oxaloacetate, glycolate and glyoxylate has been suggested by different studies. In recent years the involvement of D-erythroascorbic acid, has gained increasing attention (Franceschi and Loewus 1995).

Synthesis of Ascorbic Acid in Plants

The synthesis of ascorbic acid in plants was discussed by different authors during a relatively long time (Conklin 2001; Loewus 1999; Smirnoff et al. 2001; Smirnoff and Wheeler 2000) and was definitively clarified after the discovery of the enzyme L-galactose dehydrogenase (EC 1.1.1.316), that catalyzes the oxidation of L-galactose to L-galactono-1,4-lactone (Wheeler et al. 1998), originating the so called "Wheeler-Smirnoff" mechanism.

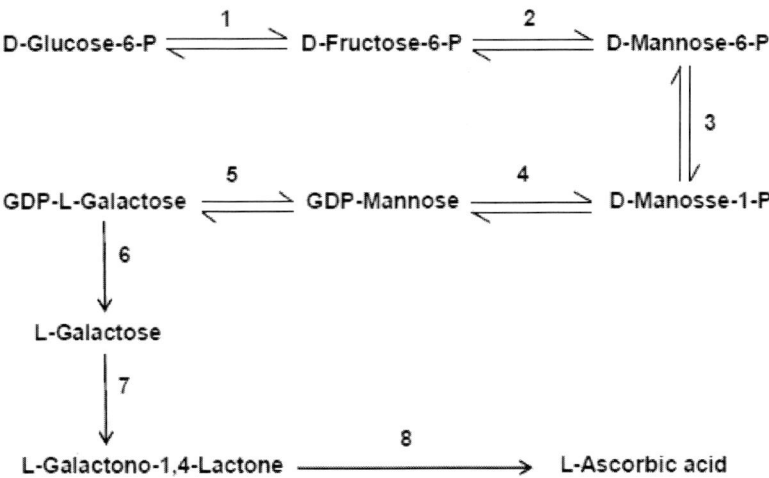

Figure 4. Schematic pathway for the biosynthesis of L-ascorbic acid in plants (Wheeler-Smirnoff mechanism). *Key to involved enzymes:* 1: D-glucose-phosphate isomerase 2: phosphomannose isomerase 3: phosphomannose mutase 4: GDP-mannose pyrophosphorylase 5: GDP-mannose-3,5-epimerase 6: activity present, but enzyme(s) uncharacterized 7: L-galactose dehydrogenase 8: mitochondrial L-galactono-1,4-lactone dehydrogenase.

In this pathway, the key intermediate for the generation of L-ascorbic acid is L-galactono-1,4-lactone. This molecule is formed, as schematized in Figure 4, by oxidation of L-galactose and which is derived from GDP-D-mannose via GDP-L-galactose. The initial observation resulting in the proposal of this pathway was that exogenous L-galactose resulted in a large increase in the total ascorbic acid pool of plant tissues similar to that caused by L-galactono-1,4-lactone. The existence of free L-galactose has never been reported in plants. It is considered to originate from GDP-D-mannose by a double epimerization (Conklin 2001; Smirnoff et al. 2001). Besides, as also shown in Figure 4, the original precursor of the synthesis is D-glucose, via D-glucose-6-phosphate.

Synthesis of Oxalic Acid in Plants

Plant related functional roles involving intact L-ascorbic acid are largely related to the redox properties associated with this molecule and this aspect of ascorbic acid metabolism has received considerable attention (Smirnoff 2000). From the chemical point of view, it is well-known that ascorbic acid is relatively unstable towards oxidation. Furthermore, dehydroascorbic acid, generated as the primary oxidation product is very unstable and undergoes a rapid series of transformations as shown schematically in Figure 5. It is degraded first to 2,3-diketogulonic acid which can further be degraded to a mixture of oxalic and L-threonic acids. At higher pH-values the latter acid is finally oxidized to tartaric acid (Davies et al. 1991). Notwithstanding, it is not clear if the mentioned intermediates are also involved in the cleavage of ascorbic acid by biological systems (Loewus 1999).

An important number of studies have shown that ascorbic acid is the major substrate for the synthesis of oxalic acid in plants. It has also been demonstrated that it derives from C1 + C2 of ascorbic acid (Kostman et al. 2001; Loewus 1999). On the other hand, L-galactose is just as effective as ascorbic acid in oxalic acid synthesis, which can be interpreted as evidence that L-galactose is converted first to ascorbic acid, in the way described above (Loewus 1999).

The C2/C3 cleavage of L-ascorbic acid produces oxalic acid, L-threonic acid and/or L-tartaric acid. The biosynthesis and possible functions of this last acid in plants is only partially understood. An alternative route for its generation may be the C4/C5 cleavage of L-ascorbic acid (Debolt et al. 2007; Loewus 1999; Smirnoff and Wheeler 2000).

Another very interesting aspect related to the generation of oxalic acid is the fact that its synthesis occurs directly within individual crystal idioblasts. Crystal idioblasts are cells that are specialized for calcium oxalate formation (Monje and Baran 2004) and they require large amounts of the acid to precipitate the excess tissue calcium they accumulate. Apparently, this cell type is self sufficient in oxalic acid biosynthesis and does not require import of oxalic acid from adjacent mesophyll cells (Franceschi and Nakata 2005; Kostman et al. 2001). Related to this is the observation that increasing the calcium level often induces an increase in the amount of crystals formed in plants. This finding suggests that the process of oxalate biosynthesis is driven or regulated by calcium levels and that crystal formation is not the end product of an attempt to sequester excess of oxalic acid, as often supposed (Franceschi and Nakata 2005).

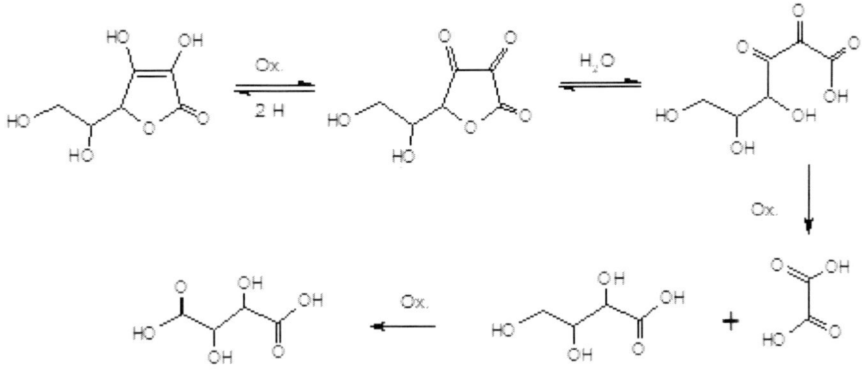

Figure 5. Schematic representation of the stepwise oxidation of L-ascorbic acid.

Synthesis of Oxalic Acid in Fungi

Reports of the presence of L-ascorbic acid, D-erithroascorbic acid, as well as other ascorbic acid analogs and their glycosides, in several fungal species are well-documented since relatively long time (Loewus 1999).

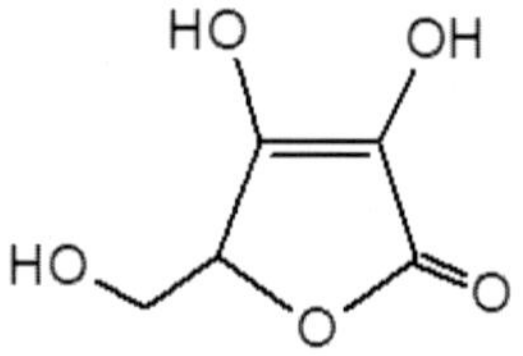

Figure 6. Schematic structure of D-erythroascorbic acid.

There are increasing evidences, that in the case of fungi, D-erythroascorbic acid is the precursor of oxalic acid (Franceschi and Loewus 1995; Loewus 1999). The enantiomeric assignment of "D" to erythroascorbic acid recovered from fungi has been confirmed by chemical synthesis and NMR analysis (Shao et al. 1993). D-erythroascorbic acid (Figure 6) differs structurally from L-ascorbic acid only in the fact that it is a five carbon rather than a six carbon compound. Ring structures are identical which leaves the side chain at C4 of D-erythroascorbic acid with a hydroxymethyl group as compared to the 1,2-dihydroxyethyl group on L-ascorbic acid. Besides, both compounds are substrates for ascorbate oxidase (Baran, 2014). Oxidation by H_2O_2 leads to oxalic acid and L-threonic acid in the case of L-ascorbic acid and to oxalic acid and D-glyceric acid in the case of D-erythroascorbic acid (Loewus, 1999).

In this case the synthetic precursor may be D-arabinose. The enzymic components of this biosynthetic pathway, from D-arabinose to D-erythroascorbic acid were originally proposed by Murakawa et al. (1977) and can be summarized in the following steps:

D-Arabinose → D-Arabino-1,5-lactone → D-Arabino-1,4-lactone → D-erythroascorbic acid

Two enzymes involved in this biosynthesis have been purified recently and cloned from *Saccharomyces cerevisiae* (Davey et al. 2000). The first step, oxidation of D-arabinose, is catalyzed by D-arabinose dehydrogenase, a NADP-dependent heterodimer which catalyzes the oxidation of D-arabinose to D-arabino-1,4-lactone. This species is then oxidized to D-erythoroascorbic acid by the flavin enzyme D-arabino-1,4-lactone oxidase (Davey et al. 2000; Loewus 1999). If D-Arabino-1,5-lactone is formed intermediately, it probably rearranges rapidly to the most stable D-Arabino-1,4-lactone form (Loewus 1999).

The analyzed pathway allows a first insight into the mechanism of the generation of the precursors of oxalic acid in fungi but, obviously, more work is needed before the role of ascorbic acid analogs is fully understood and more firmly established.

OXALATE DEGRADATION IN PLANTS AND FUNGI

General Aspects

As mentioned above, microorganisms which attack oxalate are widely distributed in Nature and it must also be remembered that oxalate degradation plays an important role in the carbon cycle (Baran and Monje 2008).

The two major classes of oxalate degrading enzymes found in plants and fungi are oxalate oxidase and oxalate decarboxylase (Baran and Monje 2008; Svedružić et al. 2005). Besides, an important number of oxalotrophic bacteria are also known (Allison et al. 1995; Baran and Monje, 2008).

Oxalate oxidase (EC 1.2.3.4) is expressed predominantly in higher plants and catalyzes the oxygen dependent oxidation of oxalate to CO_2, in a reaction that is coupled to the formation of H_2O_2 according to:

$$\text{HO-C(O)-COO}^- + O_2 \xrightarrow{H^+} 2CO_2 + H_2O_2 \tag{1}$$

Oxalate decarboxylase (EC 4.1.1.2), the second known enzyme of this type, mainly employed by fungi and some bacteria, catalyzes a remarkable reaction in which oxalate is degraded to CO_2 and formate:

$$\text{HO-C(O)-COO}^- \longrightarrow CO_2 + HCOO^- \tag{2}$$

Both enzymes share some common characteristics, among others they belong to the cupin superfamily of proteins, are manganese-dependent systems and their active sites show strong similarities.

The Cupin Superfamily

The cupin superfamily of proteins is among the most functionally diverse of any described to date. It was named on the basis of the conserved β–barrel fold ("*cupa*" is the Latin term for a small barrel), and comprises both enzymatic and non-enzymatic members, which have either one or two cupin domains. Within the conserved tertiary structure, the variety of biochemical functions is provided by minor variations of the residues in the active site and the identity of the involved metal cations. Although the majority of enzymatic cupins contains iron at the active site, other members contain copper, manganese, zinc, cobalt or nickel as a cofactor. Each of these cofactors allows, obviously, a different type of chemistry to occur within the conserved tertiary structure. It has been estimated that there is a minimum of 18 different functional subclasses of cupins, but this figure is probably much higher (Dunwell et al. 2001, 2004).

The characteristic cupin domain comprises two conserved motifs, each corresponding to two β-strands, separated by a less conserved region composed of another two β-strands with an intervening variable loop. The total size of this intermotif region varies from a minimum of 11 amino acids in some microbial enzymes, to ca. 50 amino acids in seed storage proteins, and to >100 amino acids in certain dioxygenases. Another typical structural characteristic of the family is the possibility of the existence of single cupin domains (monocupins), of a duplicated domain (bicupins) or the formation of multicupins (>2 cupin domains).

Among the monocupins the most abundant generic class comprises dioxygenases. Most of them contains iron as the active metal site and catalyzes a variety of reactions involving the oxidation of organic substrates using the O_2 molecule. A second important group is constituted by germin and germin-like proteins. With the exception of 2-oxoglutarate Fe-dependent dioxygenases, they represent the largest gene family of any cupin found in plants and the manganese-dependent oxalate oxidase belongs to this group (Dunwell et al. 2000, 2004). Examples of systems related to bicupins are quercitin dioxygenase, a copper-dependent dioxygenase that catalyses the oxidation of the plant flavonol quercitin, some iron-dependent dioxygenases involved in the degradation of different aromatic compounds and the mentioned manganese-dependent oxalate decarboxylase (Dunwell et al. 2004).

Another interesting aspect related to the cupin structure and relevant for the enzymatic-systems discussed in this review is its high thermal stability. The possible origins of this property may be related to several factors, such as important subunit contacts and efficient packing, hydrophobic interactions and hydrogen bonding, the presence of cysteine-cysteine disulfide bonds, as well as the presence of cofactors or metallic ions. An additional functional advantage provided by the compact structure of the cupin domain is its resistance to protease degradation; the lack of extensive surface loops reduces the number of sites which are accessible to enzymes capable of cleaving the protein (Dunwell et al. 2001).

Oxalate Oxidase

Oxalate oxidase (OxOx) catalyzes the oxidation of oxalate, reducing dioxygen to hydrogen peroxide and forming two moles of carbon dioxide (eq.(1), above). It is widespread in nature and has been found in a wide variety of plants, including barley seedling leaves and roots, beet stems and leaves, sorghum leaves, *Amaranthus* leaves, maize, rye, oats and rice (Svedružić et al. 2005). Finally, it was recognized that the enzyme OxOx is identical to the important seed protein germin (Lane et al. 1993). Notwithstanding, not all germins have OxOx activity, and some appear to function as manganese-dependent superoxide dismutases (Dunwell et al. 2004; Yamahara et al. 1999).

Moreover, it is important to emphasize that oxalate, like ascorbate, may participate in the oxidative biochemistry of the extracellular matrix (ECM) of higher plants. In fact, dissolution of calcium oxalate accumulated by plants, and germin-induced degradation of the resulting soluble oxalate, can release Ca^{2+} and H_2O_2, both of which are known to have central roles in the biochemistry of the ECM in higher plants (Lane 1994). On the other hand, it is also believed that the H_2O_2 generation by OxOx is employed as a defense mechanism in response to infection by pathogens (Dumas et al. 1995; Svedružić et al. 2005).

Native OxOx is a glycoprotein, the nature of the glycosylation being dependent upon the tissue from which the enzyme is obtained. The primary structures of the oxalate oxidases present in wheat, barley, maize and ryegrass have been deduced by gene cloning (Svedružić et al. 2005). A first X-ray crystallographic analysis of barley OxOx demonstrated that it crystallizes as a hexamer (a trimmer of dimmers) with a monomer molecular mass of about 26 kD (Woo et al. 1998). In a subsequent study the presence of up to 0.80 Mn(II) cation per subunit of the enzyme, was clearly demonstrated by metal analysis and EPR spectroscopy (Requena and Bornemann 1999).

Finally, a wider insight into the structural peculiarities could be obtained by a further crystallographic study at 1.6 Å resolution (Woo et al. 2000). Each OxOx monomer possesses the predicted jellyroll β-barrel that is characteristic of the cupin superfamily followed by a C-terminal domain comprised of three α-helices. The enzyme forms strongly bound dimmers, an arrangement that results in extensive burial of the monomer surfaces and hydrophobic residues. These dimmeric units then associate via their C-terminal domains to produce the hexamer, which is believed to be the biologically active form of the enzyme. This quaternary structure is probably responsible for the resistance of OxOx to degradation by heat or proteases.

The active site is located in the core of a β-barrel formed from two five-stranded β-sheets in each monomer. The Mn(II) cation is bound by the side chains of three histidines and one glutamate residue, as well as two water molecules that occupy adjacent positions in the six-coordinate distorted octahedral complex (Woo et al. 2000), as shown in Figure 7.

The main characteristics of the active metal center were also spectroscopically explored. Purified and active oxalate oxidase is colorless, although concentrated solutions have a slight yellow tinge, resulting from a weak absorption near 450 nm (Requena and Bornemann 1999; Whittaker and Whittaker 2002). It also exhibits a typical six-line ^{55}Mn (I = 5/2) mononuclear Mn(II) spectrum at g = 2. The metal is clearly protein-bound, since it does not resemble the spectrum of a $MnCl_2$ solution (Requena and Bornemann 1999; Whittaker and Whittaker 2002). This spectrum is not perturbed neither by addition of strong oxidants nor by the presence of H_2O_2. However, anaerobic addition of oxalate perturbs the spectral pattern, producing changes in the nuclear hyperfine splitting's, implying that oxalate coordinates to Mn(II) in the active site (Whittaker and Whittaker 2002). Additionally, spectroscopic information is also available for the oxidized (Mn(III)) and superoxidized (Mn(IV)) forms of oxalate oxidase (Whittaker et al. 2007).

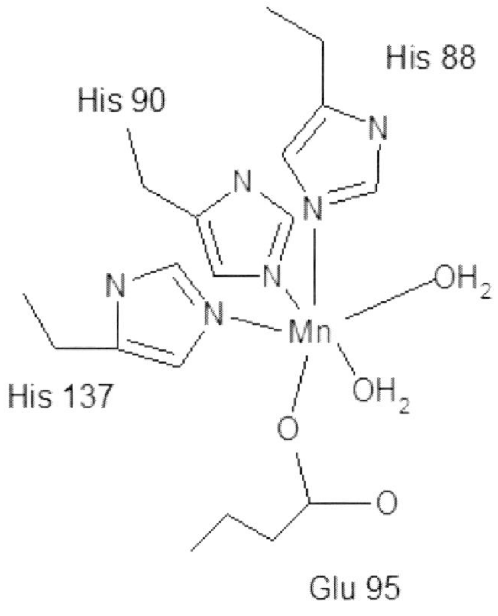

Figure 7. Schematic structure of the active site of oxalate oxidase (adapted from Woo et al. 2000).

Additionally, it is interesting to mention that a germin-like manganese superoxide dismutase (Mn-SOD) has also been described (Carter and Thornburg 2000; Yamahara et al. 1999). It is a high molecular weight glycoprotein similar to OxOx and may co-purify with oxalate oxidase from plant extracts, eventually resulting in SOD contamination of these preparations. While SOD activity has previously been reported for barley OxOx (Woo et al. 2000), the other germin-like Mn-SODs do not exhibit OxOx activity (Carter and Thornburg 2000; Yamahara et al. 1999), suggesting that the two activities may be mutually exclusive. These differences can surely be related to differences in the manganese active sites. As shown above, in OxOx manganese is six-coordinated in a distorted octahedral environment, whereas all structurally defined Mn-SODs have a five-coordinate active site of approximately trigonal bipyramidal structure (Baran 1995; Weatherburn 2001; Whittaker 2000).

Oxalate Decarboxylase

Oxalate decarboxylase (OxDc), which was initially identified in studies of the basidiomycete fungi *Collybia* (*Flammulina*) *velutipes* and *Copriolus hersutus* (Shimazono 1955; Shimazono and Hayaishi 1957; Svedružić et al. 2005), catalyzes the degradation of oxalate to formate and carbon dioxide (eq.(2), above). Subsequent studies have shown its presence in much other fungal species and for a long time it was regarded as an enzyme unique to fungi although in recent years it was also identified in different bacteria (Reinhardt et al. 2003; Svedružić et al. 2005). Moreover, an important number of studies have been precisely performed during the last years with the OxDc from one of these bacteria, *Bacillus subtilis*, a soil bacterium found in similar environments to white-rot and brown-rot fungi.

Although oxalate decarboxylases isolated from different organisms differ in some of their biochemical properties, all of these enzymes share some common features. They contain manganese and require dioxygen for catalytic turnover, even though the degradation of oxalate to CO_2 and formate involves no net redox changes. On the other hand, they exhibit optimum activity at low pH values and high specificity for oxalate as a substrate (Reinhardt et al. 2003; Tanner et al. 2001). Finally, and as commented above, they belong to the bicupin subset of the cupin superfamily. The conserved motifs appear twice in their primary sequence, presumably arising from a gene duplication event during their evolution.

Two high-resolution crystal structures have been reported for recombinant, wild type *Bacillus subtilis* oxalate decarboxylase (Anand et al. 2002; Just et al. 2004). It is a 264 kDa homohexameric enzyme made up of two trimeric layers packed face to face and having D_3 point symmetry, in which because the OxDc monomer is composed of two cupin domains (domains I and II), each trimmer resembles the OxOx hexamer. Each of the mentioned OxDc domains of the bicupin monomers has a manganese binding site that is buried deep inside the β-barrel and the two manganese sites within each monomer are separated by approximately 26 Å (Anand et al. 2002).

Interestingly, although coincident in most aspects, the two reported crystallographic structures showed significant differences between the bonding characteristics of the Mn(II) ions. In the first case, it was reported that the manganese cation in each of the domains has a distorted octahedral coordination, in which four of the ligands are contributed by highly conserved amino acid side chains of three histidines and one glutamate residue. In domain I, the remaining sites are occupied by a water molecule and a formate anion, whereas in domain II both of the remaining sites are occupied by water molecules, generating a similar complex as that depicted in Figure 7. The formate anion of domain I was present whether the enzyme was crystallized in the presence or absence of formate (Anand et al. 2002).

On the other hand, the second and most recent structural analysis revealed that a simple conformational rearrangement allows postulating the presence of totally similar Mn(II) environments in both domains, i.e., three histidines, one glutamate and two H_2O molecules, as well as a rearrangement of some of the proximal amino acid residues, which are fundamental for the understanding of the catalytic mechanism (Just et al. 2004).

Another very relevant aspect to be emphasized is the strong analogy between the cupin folds in OxDc and in oxalate oxidase, as shown by the very nice and illustrative figures presented in the paper of Anand et al. (2002). Notwithstanding, one important aspect not totally clarified is the question of which of the two metal centers present catalytic activity or if, eventually, both centers participate in the reaction. These aspects shall be analyzed and discussed in the next section.

In the case of OxDc it was not so easy as for OxOx to perform spectroscopic measurements due to the presence of two distinct Mn(II) centers. First EPR studies only confirmed the existence of a nearly octahedral geometry around these centers in the resting state, and spectral perturbations upon the addition of oxalate or formate to the enzyme. However, it was impossible to distinguish which metal signal was perturbed because the signals were very broad (Chang et al. 2004; Muthusamy et al. 2006). Most recently, a multifrequency EPR study allowed to distinguish the two Mn(II) centers, on the basis of their differing fine structure parameters, as well as to observe that acetate and formate bind to the Mn(II) cation in only one of the two sites (Angerhofer et al. 2007).

Mechanisms of Action of the Two Enzymes

Despite the detailed knowledge of the structures of oxalate oxidase and oxalate decarboxylase, the possible catalytic mechanism has not been definitively clarified, although an increasing amount of additional information has been produced in recent years for this purpose.

From the discussion of the geometries and general characteristics of the active Mn-sites present in both enzymes it is clear that the catalytic mechanisms of both enzymes must be closely related. As both enzymes use the same substrate (oxalate) it is evident that the differences in the reaction pathways and final reaction products may be related to subtle differences in the protein conformations around the active metal centers, as is also well-known for many other bioinorganic systems.

Before presenting details on the most probable catalytic mechanisms, it seems useful to make a brief survey of the most relevant available experimental data, which support these speculations:

- The optimal activity of both enzymes occurs at a pH value of about 4.0, which lies between the two pK-values for oxalic acid dissociation (1.3–4.3) (Muthusamy et al. 2006; Opaleye et al. 2006; Reinhardt et al. 2003) and, therefore, oxalic acid may be present as the monoanion, HOOC-COO$^-$. Besides, in both cases reaction occurs only in the presence of oxygen. Dioxygen is one of the substrates in the case of OxOx but acts as a cofactor in the case of OxDc.
- In the case of oxalate oxidase, in which only one manganese center is present in each protein subunit, EPR studies have conclusively shown that in the resting state this metal center is present as Mn(II) (Requena and Bornemann 1999) and it is probable that the metallic center plays not only a catalytic but also a structural role. Furthermore, all so far performed EPR studies did not give a definitive evidence for the presence of Mn(III) during enzymatic turnover suggesting that this oxidation state is only transiently formed and rapidly reduced to Mn(II).
- In contrast to Mn-dependent superoxide dismutase, which binding site stabilized Mn(III) (Whittaker 2000), both OxOx and OxDc favor Mn(II) in its resting state making it necessary to oxidize the metal in order to generate catalytic activity. It can be assumed that oxalate binding sufficiently perturbs the redox potential so as to favor the transient formation of Mn(III) by dioxygen in a ternary Enzyme-oxalate-O$_2$ complex. To gain further insight into the reaction mechanisms it may be useful to know the Mn(III)/Mn(II) reduction potential. However, no experimental data are available for the enzymatic systems although it was recently suggested that in the case of OxOx, and based on calculations of the free energy change for the oxidative cleavage of oxalate, this reduction potential must lie in the range +0.4 to +1.0 V (vs. NHE)

(Whittaker et al. 2007). These values are biologically reasonable, as they are less than E° for the Mn(III)/Mn(II) redox couple of the free metal ion (+1.51 V) (Porterfield 1993).
- A structural investigation of substrate binding on OxOx, using glycolate as a structural analogue for oxalate, shows monodentate binding to the metal center (Opaleye et al. 2006).
- In the case of oxalate decarboxylase, in which two Mn(II) centers are present in each bicupin subunit, recent EPR studies (Angerhofer et al. 2007) clearly demonstrate that the metal cation located in domain I is the catalytic active center. Then, the second Mn-center surely plays only a structural role (Just et al. 2004).

In spite of the fact that a number of questions related to the mechanism of action of these enzymes remains to be explored to attain a definitive and solid model for these processes, on the basis of the so far accumulated information it is possible to present a plausible general catalytic mechanism for both discussed systems. As commented above, and taking into account the similar structures of the active centers, it is evident that a number of common mechanistic aspects must be shared by both enzymes.

The mechanisms compatible with all the so far accumulated information are presented in Figure 8, and are briefly commented as follows.

For clarity, the four permanent (N_3O) ligand atoms bonded to the metal center are omitted.

- During the full reaction, the manganese center remains bound to the four conserved protein residues (three histidines and one glutamate). Only the two H_2O molecules are displaced in the initial reaction step. In this first step, mono-protonated oxalate and dioxygen bind to the Mn(II) center. Oxidation of this site to Mn(III) is then thought to take place with concomitant formation of a Mn-bound superoxide anion.

Oxalate Biosynthesis and Degradation in Plants and Fungi 115

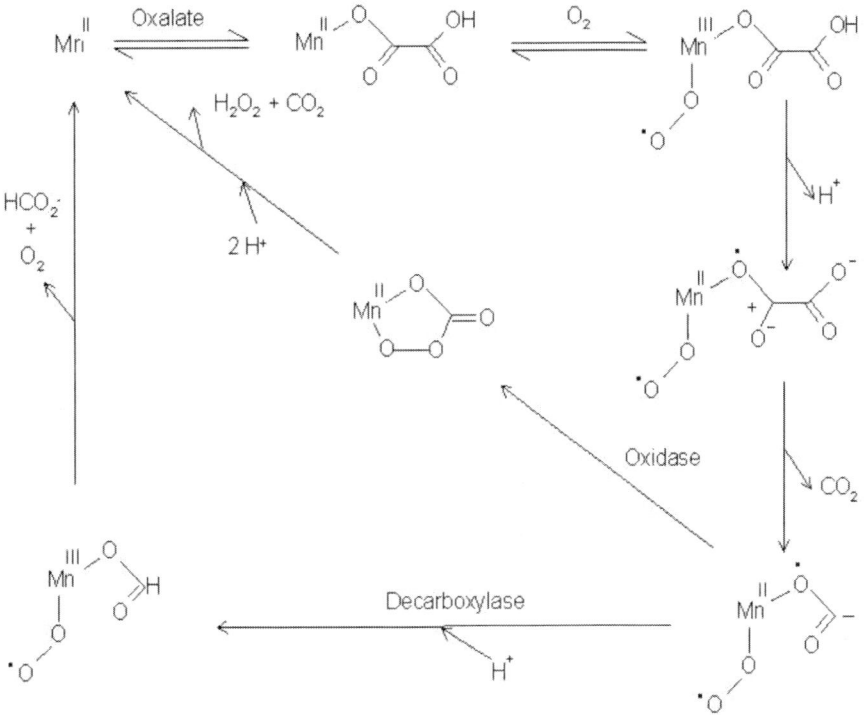

Figure 8. Proposed catalytic mechanism for oxalate oxidase and oxalate decarboxylase (adapted from Just et al. 2004; Burrell et al. 2007).

- An active site residue, probably a glutamate (Svedružić et al. 2007), can then mediate proton-coupled electron transfer in the initial complex to give a Mn(II)-bound oxalate radical anion.
- This anion undergoes heterolytic bond cleavage to release CO_2, generating a Mn(II) bound formate radical anion.
- This intermediate species is common to the two enzymes and its subsequent transformation depends on the protein residues located near the active metal center.
- In the case of oxalate decarboxylase, the formate radical is protonated with participation of a glutamate residue, again with electron transfer from the metal, yielding Mn(III) bound formate. The glutamate residue Glu162 is apparently crucial for the concretion of this reaction as its presence differentiates the two

enzymes, generating the ability of the decarboxylase to specifically protonate the carbon atom of this intermediate. (Burrell et al. 2007; Just et al. 2004; Muthusamy et al. 2006; Svedružić et al. 2007).
- As shown by recent FTIR studies, the source of this proton is the solvent (Muthusamy et al., 2006).
- In the final step of the OxDc reaction, formate and dioxygen are released, with the concomitant reduction of the active center to Mn(II).
- In the case of oxalate oxidase, the intermediate species suffers a rearrangement generating a bidentate peroxocarbonate complex, followed by the loss of a second CO_2 molecule, coupled with the protonation of the peroxide group and release of H_2O_2.

Some aspects of the presented mechanism deserve some interesting additional comments. From the chemical point of view the intermediate generation of a peroxocarbonate complex in the case of OxOx is somewhat unexpected but attractive. A revision of the literature shows that the existence of peroxocarbonate complexes is not new and they have reported especially for some heavy transition metals, such as Pt, Rh, Ir or Pd (Aresta et al. 1996; Dahlenburg and Prengel 1984; Hayward et al. 1970). They have been usually prepared by reaction of dioxygen complexes with carbon dioxide, or inversely treating carbon dioxide complexes with dioxygen (Aresta et al. 1996). Most recently, the mode of peroxocarbonate binding to the Mn(II) center has also been reported for an Fe(III) complex of composition $[Fe(qn)_2(O-OC(O)O]^-$ (qn = quinaldinate, each of which provides a N and an O ligand), (Furutachi et al. 2005; Hashimoto et al. 2002). Interestingly, it is also reported that the epoxidation of alkenes can be efficiently produced using H_2O_2 in the presence of $MnSO_4$ in a bicarbonate buffer, and it is assumed that the catalytic mechanism involves a Mn(II) peroxocarbonate complex (Lane et al. 2002). Besides, it was also speculated that this active catalytic species may be generated by a direct interaction of Mn(II) with peroxomonocarbonate (generated by the reaction $HCO_3^- + H_2O_2 \rightarrow HCO_4^- + H_2O$) or by the initial formation of a peroxocomplex (Mn-O-OH) and its subsequent reaction with HCO_3^- (Lane et al. 2002).

On the other hand, it is also valuable to comment that the catalytic reaction mechanism of oxalate oxidase has been theoretically investigated by means of a DFT approach (Borowski et al. 2005). This study basically supports the mechanism presented in Figure 8 and discussed above, and also allows to discard other, somewhat different, mechanistic proposals (Whittaker and Whittaker, 2002). These calculations also confirm the central role of manganese in this enzyme, as the metal center is involved in the following processes: a) it brings the substrates together, b) temporarily provides the one electron necessary for the dioxygen reduction, c) facilitates the required fast spin transition (from a quartet to a sextet spin state) through an efficient spin-orbit coupling route (Borowski et al. 2005).

Finally, it is also interesting to comment that the oxygen activation chemistry discussed here for the two oxalate degrading enzymes has precedents in a distinct class of Mn-metalloenzymes, the manganese-dependent catechol dioxygenases, which also share significant structural similarities. They contain a mononuclear Mn(II) center coordinated by two N-atoms of histidine residues, one O-atom of a glutamate moiety and two water molecules. What is more, in these enzymes aromatic ring-cleavage is thought to be driven by O_2 attack on a monodentate coordinated substrate to the metal center (Que and Reynolds 2000).

OXALOTROPHIC BACTERIA

Oxalatrophy refers to the ability of an organism to use oxalic acid or oxalates as carbon and electron sources. Oxalates and oxalic acid are found in a wide range of environments and at different concentrations and oxalatrophy is mainly related to bacterial catabolism (Hervé et al. 2016).

More than 50 types of oxalotrophic bacteria have been identified to date (Baran and Monje 2008; Sahin 2003; Svedružić et al. 2005). Some of these bacteria are "generalists," because they can ferment many other substrates apart from oxalate. Others are "specialists" because they only use oxalate as a carbon and energy source (Allison et al. 1995). Notwithstanding, oxalate degrading enzymes have been isolated and characterized from only a small number of microorganisms.

Two key enzymes enzymes are involved in bacterial oxalate degradation, i.e., oxalyl-CoA-decarboxylase (EC 4.1.1.8), a thiamine diphosphate (ThDP) dependent enzyme and formyl-CoA-transferase (EC 2.8.3.16) (Baran and Monje, 2008; Hervé et al. 2016; Svedružić et al. 2005). The first one catalyzes the decarboxylation of oxalyl-CoA to generate CO_2 and formyl-CoA (equation (3)). On the other hand, formyl-CoA-transferase is used to activate oxalate to oxalyl-CoA before this decarboxylation takes place (Baetz and Allison 1989, 1990; Svedružić et al. 2005) (equation (4)). The overall catalytic cycle therefore converts oxalate to formate and carbon dioxide with consumption of a proton.

$$\text{}^{-}\text{OO-C-C(O)-CoA} \xrightarrow[\text{H}^+]{\text{ThDP}} \text{HC(O)-CoA} + CO_2 \qquad (3)$$

$$\text{HC(O)-CoA} + {}^{-}\text{OO-C-COO}^{-} \longrightarrow \text{HCOO}^{-} + {}^{-}\text{OO-C-C(O)-CoA} \qquad (4)$$

Aerobic as well as anaerobic oxalate-degrading bacteria are known. Information about oxalate degradation by anaerobic microbes comes mainly from studies of activities in the rumen and the mammalian hind gut. Aerobic bacteria have often been found in terrestrial ecosystems (Allison et al. 1995).

Oxalate is commonly found in soil and is primarily derived from root exudates, breakdown products from plants, animal and microbial tissues, and metabolites from bacteria and fungi. The oxalate concentration in soils reflects seasonal cycles and oxalate in soils is constantly being synthesized and degraded. Due to its capacity to chelate metals, oxalate plays a central role in the solubilization and transport of soil metals and to the weathering of rocks (Adamo and Violante 2000; Hofmann and Bernasconi 1998). Through interactions with aluminum and iron, oxalate plays a major role in plant nutrition by increasing the availability of phosphorus, potassium, magnesium and calcium in soils. Besides, oxalate is also involved in the detoxification of aluminum and calcium by removing these cations from soil solutions. Therefore, aerobic or anaerobic microbial processes that affect oxalate-metal interactions or the availability of oxalate may ultimately influence the nutritional-toxicological status of terrestrial ecosystems and has a high impact on plant physiology (Allison et al. 1995; Sahin 2003).

ACKNOWLEDGMENTS

The author is an Emeritus Professor from the Universidad Nacional de La Plata and the continuous support from this University is gratefully acknowledged.

REFERENCES

Adamo, P. and Violante, P. (2000). Weathering of rocks and neogenesis of minerals associated with lichen activity. *Applied Clay Science*, 16, 229-256.

Allison, M. J., Daniel, S. L. and Cornick, N. A. (1995). Oxalate degrading bacteria. In: S. R. Khan (Ed.) *Calcium Oxalate in Biological Systems* (pp. 131-168). Boca Raton: CRC-Press.

Anand, R., Dorrestein, P. C., Kinsland, C., Begley, T. B. and Ealick, S. E. (2002). Structure of oxalate decarboxylase from *Bacillus subtilis* at 1.75 Å resolution. *Biochemistry*, 41, 7659-7669.

Angerhofer, A., Moomaw, E. W., García-Rubio, I., Ozarowski, A., Krzystek, J., Weber, R. T. and N. G. J. Richards (2007). Multifrequency EPR studies on the Mn(II) centers of oxalate decarboxylase. *Journal of Physical Chemistry B*, 111, 5043-5046.

Aresta, M., Tommasi, I., Quarante, E., Fragale, C., Mascetti, J. Tranquille, M., Galan, F. and Fouassier, M. (1996). Mechanism of formation of peroxocarbonates $RhOOC(O)O(Cl)(P)_3$ and their reactivity as oxygen transfer agents mimicking monooxygenases. The first evidence of CO_2 insertion into the O-O bond of Rh ($^2\eta$-O_2) complexes. *Inorganic Chemistry*, 35, 4254-4260.

Arnott, H. J. (1982). Three systems of biomineralization in plants with comments on the associated organic matrix. In: G. H. Nancollas (Ed.) *Biological Mineralization and Demineralization* (pp. 199-218). Berlin: Springer.

Baetz, A. L. and Allison, M. J (1989). Purification and characterization of oxalyl-conzyme A decarboxylase from *Oxalobacter formigenes*. *Journal of Bacteriology*, 171, 2605-2608.

Baetz, A. L. and Allison, M. J. (1990). Purification and characterization of formyl-coenzyme A transferase from *Oxalobacter formigenes*. *Journal of Bacterioliology*, 172, 3577-3540.

Baran, E. J. (1995). *Química Bioinorgánica*. Madrid: McGraw-Hill Interamericana de España S. A. [*Inorganic Biochemistry*. Madrid: McGraw-Hill Interamericana de España S. A].

Baran, E. J. (2014). Copper in plants: An essential and multifunctional element. In: A. Hemantaranjan (Ed.) *Advances in Plant Physiology* (Vol. 14, pp. 373-397). Jodhpur: Scientific Publishers.

Baran, E. J. (2016). Natural iron oxalates and their analogous synthetic counterparts: A review. *Chemie der Erde – Geochemistry*, 76, 449-460.

Baran, E. J. and Monje, P. V. (2008). Oxalate biominerals. In: A. Sigel, H. Sigel and R. K. O. Sigel (Eds.) *Metal Ions in Life Sciences* (Vol. 4, pp. 219-254). Chichester: Wiley.

Borowski, T., Bassan, A., Richards, N. G. J and Siegbahn, P. E. M. (2005). Catalytic reaction mechanism of oxalate oxidase (Germin). A hybrid DFT study. *Journal of Chemical Theory and Computation*, 1, 686-693.

Burrell, M. R., Just, V. J., Bowater, L., Fairhust, S. A., Requena, L., Lawson, D. M. and Bornemann, S. (2007). Oxalate decarboxylase and oxalate oxidase activities can be interchanged with a specificity of up to 282000 mutating an active site lid. *Biochemistry*, 46, 12327-12336.

Carter. C. and Thornburg, R. W. (2000). Tobacco Nectarin I. *Journal of Biological Chemistry*, 275, 36726-36733.

Chang, C. H., Svedružić, D., Ozarowski, A., Walker, L., Yeagle, G., Britt, R. D., Angerhofer, A. and Richards, N. G. J. (2004). EPR spectroscopic characterization of the manganese center and a free radical in the oxalate decarboxylase reaction. *Journal of Biological Chemistry*, 279, 52840-52849.

Conklin, P. L. (2001). Recent advances in the role and biosynthesis of ascorbic acid in plants. *Plant, Cell & Environment*, 24, 383-394.

Dahlenburg, L. and Prengel, C. (1984). Alkyl and aryl compounds of iridium and rhodium. 18. Oligophosphine ligands. 6. Reactivity of some alkyls and aryls of rhodium and iridium toward carbon dioxide. Facile formation and X-ray structural characterization of the peroxocarbonato complex [cyclic]-*mer*-Rh(4-MeC$_6$H$_4$) [OOC(O)][PhP(CH2CH2CH2PPh2)2]. *Organometallics*, 3, 934-936.

Darken, L. S. (1941). The ionization constants of oxalic acid at 25 ° from conductance measurements. *Journal of the American Chemical Society*, 63, 1007-1011.

Davey, M. W., van Montagu, M., Inzé, D., Sanmartin, M., Kanellis, A., Smirnoff, N., Benzie, I. J. J., Strain, J. J., Falvell, D. and Fletcher, J. (2000). Plant L-ascorbic acid: Chemistry, function, metabolism, bioavailability and effects of processing. *Journal of the Science of Food and Agriculture*, 80, 825-860.

Davies, M. B., Austin, J., Partridge, D. A. (1991). *Vitamin C: Its Chemistry and Biochemistry*. London: Royal Society of Chemistry.

Debolt, S., Melino, V. and Ford, C. M. (2007). Ascorbate as a biosynthetic precursor in plants. *Annals of Botany*, 99, 3-8.

Deganello, S., Kampff, A. R. and Moore, P. B. (1981). The crystal structure of calcium oxalate trihydrate: Ca(H$_2$O)$_3$(C$_2$O$_4$). *American Mineralogist*, 66, 859-865.

Dumas. B., Freyssinet, G. and Pallet, K. E. (1995). Tissue-specific expression of germin-like oxalate oxidase during development and fungal infection of barley seedlings. *Plant Physiology*, 107, 1091-1096.

Dunwell, J. M., Culham, A., Carter, C. E., Sosa-Aguirre, C. R., Goodenough, P. W. (2001). Evolution of functional diversity in the cupin superfamily. *Trends in Biochemical Sciences*, 26: 740-746.

Dunwell, J. M., Khuri, S. and Gane, P. J. (2000). Microbial relatives of the seed storage proteins of higher plants: Conservation of the structure and diversification of function during evolution of the cupin superfamily. *Microbiology and Molecular Biology Reviews*, 64, 153-179.

Dunwell, J. M., Purvis, A. and Khuri, S. (2004). Cupins: The most functionally diverse protein superfamily? *Phytochemistry*, 65, 7-17.

Echigo, T. and Kimata, M. (2010). Crystal chemistry and genesis of organic minerals: a review of oxalate and polycyclic aromatic hydrocarbon minerals. *Canadian Mineralogist*, 48, 1329-58.

Franceschi, V. R. and Loewus, F. A. (1995). Oxalate biosynthesis and functions in plants and fungi. In: S. R. Khan (Ed.) *Calcium Oxalate in Biological Systems* (pp. 113-130). Boca Raton: CRC-Press.

Franceschi, V. R. and Nakata, P. A. (2005). Calcium oxalate in plants: Formation and function. *Annual Reviews in Plant Biology*, 56, 41-71.

Furutachi, H., Hashimoto, K., Nagatomo, S., Endo, T., Fujinami, S., Watanabe, Y., Kitagawa, T. and Suzuki, M. (2005). Reversible O-O bond cleavage and formation of a peroxo moiety of a peroxocarbonate ligand mediated by an iron (III) complex. *Journal of the American Chemical Society*, 127, 4550-4551.

Hashimoto, K., Nagatomo, S., Fujinami, S., Furutachi, H., Ogo, S., Suzuki, M., Uehara, A., Maeda, Y., Watanabe, Y. and Kitagawa, T (2002). A new mononuclear iron(III) complex containing a peroxocarbonate ligand. *Angewandte Chemie International Edition*, 41, 1202-1205.

Hayward, P. J., Blake, D. M., Wilkinson, G., Nyman, C. J. (1970). Some reactions of peroxobis(triphenylphosphine)platinum(II) and analogs with carbon dioxide, carbon disulfide, and other unsaturated molecules. *Journal of the American Chemical Society*, 92, 5873-5878.

He, H., Veneklaas, E. J., Kuo, J. and Lambers, H. (2014). Physiological and ecological significance of biomineralization in plants. *Trends in Plant Science*, 19, 166-174.

Hervé, V., Junier, T., Bindschedler, S., Verrecchia, E. and Junier, P. (2016). Diversity and ecology of oxalotrophic bacteria. *World Journal of Microbiology and Biotechnology*, 32, art. 28.

Hofmann, B. A. and Bernasconi, S. M. (1998). Review of occurrence and carbon isotope geochemistry of oxalate minerals: Implications for the origin and fate of oxalate in diagenetic and hydrothermal fluids. *Chemical Geology*, 149, 127-146.

Just, V. J., Stevenson, C. E. M., Bowater, L., Tanner, A., Lawson, D. M. and Bornemann, S. (2004). A closed conformation of *Bacillus subtilis* oxalate decarboxylase provides evidence for the true identity of the active site. *Journal of Biological Chemistry*, 279, 19867-19874.

Khan, S. R. (Ed.) (1995). *Calcium Oxalate in Biological Systems*. Boca Raton: CRC-Press.

Kostman, T. A., Tarlyn, N. M., Loewus, F. A. and Franceschi, V. R. (2001). Biosynthesis of L-ascorbic acid and conversion of carbons 1 and 2 of L-ascorbic acid to oxalic acid occurs within individual calcium oxalate crystal idioblasts. *Plant Physiology*, 125, 634-640.

Lane, B. G., Dunwell, J. M., Ray, J. A., Schmitt, M. R. and Cuming, A. C. (1993). Germin, a protein marker of early plant development, is an oxalate oxidase. *Journal of Biological Chemistry*, 268, 12239-12242.

Lane, B. G. (1994). Oxalate, germin, and the extracellular matrix of higher plants. *FASEB Journal*, 8, 294-301.

Lane, B. S., Vogt, M., DeRose, V. J. and Burgess, K. (2002). Manganese-catalyzed epoxidations of alkenes in bicarbonate solutions. *Journal of the American Chemical Society*, 124, 11946-11954.

Libert, B. and Franceschi, V. R. (1987). Oxalate in crop plants. *Journal of Agricultural and Food Chemistry*, 35, 926-938.

Loewus, F. A. (1999). Biosynthesis and metabolism of ascorbic acid in plants and of analogs of ascorbic acid in fungi. *Phytochemistry*, 52, 193-210.

Monje, P. V. and Baran, E. J. (2004). Plant biomineralization. In: A. Hemantaranjan (Ed.) *Advances in Plant Physiology* (Vol. 7, pp. 395-410). Jodhpur: Scientific Publishers.

Murakawa, S., Sano, S., Yamashita, H. and Takahashi, T. (1977). Biosynthesis of D-erythroascorbic acid by *Candida*. *Agricultural and Biological Chemistry*, 41, 1799-1800.

Muthusamy, M., Burrell, M. R., Thomeley, R. N. F. and Bornemann, S. (2006). Real-time monitoring of the oxalate decarboxylase reaction and probing hydron exchange in the product, formate, using Fourier transform infrared spectroscopy. *Biochemistry*, 45, 10667-10673.

Nakata, P. (2003). Advances in our understanding of calcium oxalate crystal formation and function in plants. *Plant Science*, 164, 901-909.

Opaleye, O., Rose, R. S., Whittaker, M. M., Woo, E-J-., Whittaker, J. W. and Pickersgill, R. W. (2006). Structural and spectroscopic studies shed light on the mechanism of oxalate oxidase. *Journal of Biological Chemistry*, 281, 6428-6433.

Porterfield, W. W. (1993). *Inorganic Chemistry. A Unified Approach*, 2nd. Edit. San Diego: Academic Press.

Que, L. and Reynolds, M. F. (2000). Manganese(II)-dependent extradiol-cleaving catechol dioxygenases. In: A. Sigel and H. Sigel (Eds.) *Metal Ions in Biological Systems* (Vol. 37, pp. 505-525). New York: Marcel Dekker.

Raven, J. A., Griffith, H., Glidewell, S. M. and Preston, T. (1982). The mechanism of oxalate biosynthesis in higher plants: Investigations with the stable isotopes ^{18}O and ^{13}C. *Proceedings of the Royal Society, London B*, 216, 87-101.

Reinhardt, L. A., Svedružić, D., Chang, C. H., Cleland, W. W., Richards, N. G. J. (2003). Heavy atom isotope effects on the reaction catalyzed by the oxalate decarboxylase from *Bacillus subtilis*. *Journal of the American Chemical Society*, 125, 1244-1252.

Requena, L. and Bornemann, S. (1999). Barley (*Hordeum vulgare*) oxalate oxidase is a manganese-containing enzyme. *Biochemical Journal*, 343, 185-190.

Sahin, N. (2003). Oxalotrophic bacteria. *Research in Microbiology*, 154, 399-407.

Shao, Y. Y., Seib, P. A., Kramer, K. J. and van Galen, D. A. (1993). Synthesis and properties of D- erythroascorbic acid and its vitamin C activity in the tobacco hornworm (*Manduca sexta*). *Journal of Agricultural and Food Chemistry*, 41, 1391-1396.

Shimazono, H. (1955). Oxalic acid decarboxylase, a new enzyme from the mycelium of wood destroying fungi. *Journal of Biochemistry (Tokio)*, 42, 321-340.

Shimazono, H. and Hayaishi, O. (1957). Enzymatic decarboxylation of oxalic acid. *Journal of Biological Chemistry*, 227, 151-159.

Smirnoff, N. (2000). Ascorbic acid: metabolism and functions of a multi-facetted molecule. *Current Opinion in Plant Biology*, 3, 229-235.

Smirnoff, N. and Wheeler, G. L. (2000). Ascorbic acid in plants: Biosynthesis and function. *Critical Reviews in Biochemistry and Molecular Biology*, 35, 291-314.

Smirnoff, N., Conklin, P. L. and Loewus, F. A. (2001). Biosynthesis of ascorbic acid in plants: A renaissance. *Annual Review of Plant Physiology and Plant Molecular Biology*, 52, 437-467.

Strunz, H. and Nickel, E. H. (2001). *Strunz Mineralogical Tables*, 9th Edition, Stuttgart: E. Schweizerbart'sche Verlagsbuchhandlung.

Svedružić, D., Jónsson, S., Toyota, C. G., Reinhardt, L. A., Ricagno, S., Lindqvist, Y. and Richards, N. G. J. (2005). The enzymes of oxalate metabolism: unexpected structures and mechanisms. *Archives of Biochemistry and Biophysics*, 433, 176-192.

Svedružić, D., Liu, W., Reinhardt, L. A., Wroclawska, E., Cleland, W. W. and Richards, N. G. J (2007). Investigating the roles of putative active site residues in the oxalate decarboxylase from *Bacillus subtilis*. *Archives of Biochemistry and Biophysics*, 464, 36-47.

Tanner, A., Bowater, L., Fairhurst, S. A. and Bornemann, S. (2001). Oxalate decarboxylase requires manganese and dioxygen for activity. *Journal of Biological Chemistry*, 276, 43627-43634.

Veis, A. (2008). Crystals and life: An introduction. In: A. Sigel, H. Sigel and R. K. O. Sigel (Eds.) *Metal Ions in Life Sciences* (Vol. 4, pp. 1-35). Chichester: Wiley.

Weatherburn, D. C. (2001). Manganese-containing enzymes and proteins. In: I. Bertini, A. Sigel and H. Sigel (Eds.) *Handbook of Metalloproteins* (pp. 193-168). New York: Marcel Dekker.

Webb, M. A (1999). Cell-mediated crystallization of calcium oxalate in plants. *Plant Cell*, 11, 751-761.

Weiner, S. and Dove, P. M. (2003). An overview of biomineralization processes and the problem of the vital effect. In: P. M. Dove, J. J. De Yoreo and S. Weiner (Eds.) *Reviews in Mineralogy and Geochemistry: Biomineralization* (Vol. 54, pp. 1-19). Washington D. C.: Mineralogical Soc./Geochem. Soc.

Wheeler, G. L., Jones, M. A. and Smirnoff, N. (1998). The biosynthetic pathway of vitamin C in higher plants. *Nature*, 393, 365-368.

Whittaker, J. W. (2000). Manganese superoxide dismutase. In: A. Sigel and H. Sigel (Eds.) *Metal Ions in Biological Systems* (Vol. 37, pp. 587-611). New York: Marcel Dekker.

Whittaker MM, Whittaker JW (2002) Characterization of recombinant barley oxalate oxidase expressed by *Pichia pastoris*. J. Biol. Inorg. Chem. 7: 136-145.

Whittaker, M. M., Pan, H. Y., Yukl, E. T. and Whittaker, J. W. (2007). Burst kinetics and redox transformations of the active site manganese ion in oxalate oxidase. *Journal of Biological Chemistry*, 282, 7011-7023.

Woo, E. J., Dunwell, J. M., Goodenough, P. W. and Pickersgill, R. W. (1998). Barley oxalate oxidase is a hexameric protein related to seed storage proteins: evidence from X-ray crystallography. *FEBS Letters*, 437, 87-90.

Woo, E. J., Dunwell, J. M., Goodenough, P. W., Marvier, A. C. and Pickersgill, R. W. (2000). Germin is a manganese containing hexamer with oxalate oxidase and superoxide dismutase activities. *Nature Structural Biology*, 7, 1036-1040.

Yamahara, T., Shiono, T., Suzuki, T., Tanaka, K., Takio, S., Sato, K., Yamazaki, S. and Satoh, T. (1999). Isolation of a germin-like protein with manganese superoxide dismutase activity from cells of a moss, *Barbula unguiculata*. *Journal of Biological Chemistry*, 274, 33274-33278.

Young, R. A. and Brown, W. E. (1982). Structures of biological minerals. In: G. H. Nancollas (Ed.) *Biological Mineralization and Demineralization* (pp. 101-141). Berlin: Springer.

BIOGRAPHICAL SKETCH

Enrique J. Baran was born in Olavarría (province of Buenos Aires, Argentina) in 1940. He received his Doctorate in Chemistry (PhD) from the National University of La Plata (UNLP), Argentina, in 1967, in the field of Inorganic Chemistry. He made post-doctoral research at the Institute of Inorganic Chemistry, University of Göttingen (1968-70) and at the Faculty of Chemistry, University of Dortmund (1974), in both opportunities as an "Alexander von Humboldt" fellow and under the supervision of Prof. Achim Müller. Full Professor of Inorganic Chemistry (1981-2005) and Emeritus Professor of the UNLP (2008). Honorary Professor of the Universidad Nacional de Tucumán (2013). Research Fellow from the National Research Council, Consejo Nacional de Investigaciones Científicas y Técnicas-CONICET (1970-2012). Retired as "Investigador Superior" (highest category in this career). Visitant Professor: Universities of Colombia, Germany, Spain and Uruguay. Director of the Center of Inorganic Chemistry-CEQUINOR (2001-2006). Member of the National Academy of Exact, Physical and Natural Sciences (1996-cont.; General Secretary of the Academy since 2006) and of TWAS (1997-cont.). National Representative at IUPAC (1984-1992). Author of the first text book on Inorganic Biochemistry, in Spanish language (McGraw-Hill, 1995) and of more than 700 scientific publications. Member of the Editorial Boards of the *Journal of Inorganic Biochemistry* (1992-1996), *Latin American Journal of Pharmacy* (1992-cont.), *Industria & Química* (2007-cont.), *Journal of Coordination Chemistry* (2008-2012), *Biological Trace Element Research* (2009-cont.) and *Advances in Plant Physiology* (2013-cont.). Main research interests: Coordination chemistry, solid state chemistry, biomineralization, vibrational spectroscopy, Bioinorganic Chemistry, Medicinal Inorganic Chemistry.

Address: Centro de Química Inorgánica –CEQUINOR- Facultad de Ciencias Exactas, Universidad Nacional de La Plata, Bvd. 120, Nr. 1465, 1900-La Plata, ARGENTINA.

Honors:
- "Rafael A. Labriola Prize" for distinguished young scientists (Chemical Society of Argentina, 1982).
- "Platinum Konex-Prize" as the most relevant figure in the areas of Physical Chemistry and Inorganic Chemistry in the 1983/92 decade (Konex Foundation, 1993).
- "Hans J. Schumacher-Prize" of the National Academy of Exact, Physical and Natural Sciences (1993).
- TWAS-Award in Chemistry (TWAS, Trieste, 1996).
- "Cincuentenario"-Prize (Argentine Association for the Advancement of Science, 1997).
- "Horacio Damianovich-Prize" (Consecration Prize in Inorganic Chemistry from the Chemical Society of Argentina, 2004).
- Edition of a special issue of the *Journal of the Argentine Chemical Society* (vol. 97(1) (2009)) in his homage.
- Designation as Distinguished Graduate (National University of La Plata, 2010).
- Honorary Professor of the Universidad Nacional de Tucumán (Tucumán, 2013).

Publications from the Last 3 Years:

Regular papers:
1. Baran, E. J.,: "Mean Amplitudes of Vibration of the $VO_2F_2^-$ and $VO_2Cl_2^-$ Anions" *Phys. Chem., An Indian J*, 12, 1-5 (2017).
2. Echeverría, G. A., O. E. Piro, J. Zinczuk & E. J. Baran: "Three New Thiosaccharinate Derivatives Generated in a Complex Reaction System," *J. Argent. Chem. Soc.* 104, 1-10 (2017).
3. León, I. E., P. Diez, E. J. Baran, S. B. Etcheverry & M. Fuentes: "Decoding the Anticancer Activity of VO-Clioquinol Compounds:

The Mechanism of Action and Cell Death Pathway in Human Osteosarcoma Cells," *Metallomics* 9, 891-901 (2017).
4. Echeverría, G. A., O. E. Piro, B. S. Parajón-Costa & E. J. Baran: "Structural and IR-Spectroscopic Characterization of Cadmium and Lead(II) Acesulfamates," *Z. Naturforsch.* 72b, 739-745 (2017).
5. Piro, O. E., G. A. Echeverria, B. S. Parajón-Costa & E. J. Baran: "Structural and IR-Spectroscopic Characterization of Aqua Lithium Acesulfamate, an Outlier of the M(ace), M: Na^+, K^+, Rb^+, Cs^+, Isomorphic Series," *J. Chem. Crystallogr.* 47, 226-232 (2017).
6. Baran, E. J.,: "Terapia por Captura de Neutrones," *Revista de ADEQ* 3, 12-16 (2017). [Neutron Capture Therapy. *ADEQ Journal* (Montevideo) 3, 12-16 (2017)].
7. Baran, E. J.: "La Química Bioinorgánica en el Contexto de un Curso Moderno de Química Inorgánica," *Industria y Química* 368, 31-35 (2017). [Bioinorganic Chemistry in the Context of a Modern Course of Inorganic Chemistry. *Chemistry and Industry* 368, 31-35 (2017)].
8. Baran, E. J.: "Vanadio: Un Nuevo Elemento Estratégico ?," *Anales Acad. Nac. Cs. Ex. Fís. y Nat.* 69, 84-114 (2017). [Vanadium: A New Strategic Element?. *Annals of the National Academy of Exact, Physical and Natural Sciences* 69, 84-114 (2017)].
9. Piro, O. E., G. A. Echeverria, A. C. González-Baró & E. J. Baran: "Crystal Structure and Spectroscopic Behavior of Synthetic Novgorodovaite $Ca_2(C_2O_4)Cl_2 \cdot 2H_2O$ and its Twinned Triclinic Heptahydrate Analog," *Phys. Chem. Min.* 45, 185-195 (2018).
10. Piro, O. E., & E. J. Baran: "Crystal Chemistry of Organic Minerals - Salts of Organic Acids: The Synthetic Approach," *Crystallogr. Reviews* 24, 149-175 (2018).
11. Piro, O. E., G. A. Echeverría & E. J. Baran: "Spontaneous Enantiomorphism in Poly-phased Alkaline Salts of *tris*(oxalato)ferrate(III): Crystal Structure of Cubic $NaRb_5[Fe(C_2O_4)_3]_2$," *Acta Crystallogr.* E74, 905-909 (2018).
12. D'Antonio, M. C., M. M. Torres, D. Palacios, A. C. González-Baró, V. L. Barone & E. J. Baran: "Structural and Spectroscopic Behavior

of Double Metal Oxalates from the First Transition Metal Series," *Anales Asoc. Quím. Argent.* 105, 41-48 (2018).
13. Baran, E. J., O. E. Piro, G. A. Echeverría & B. S. Parajón-Costa: "Structural and IR-Spectroscopic Characterization of Pyridinium Acesulfamate, a Monoclinic Twin," *Z. Naturforsch.* 73 b, 753-758 (2018).
14. Baran, E. J.: "Cobalto: Un Elemento Crítico y Estratégico," *Anales Acad. Nac. Cs. Ex. Fís. y Nat.* 70, 77-106 (2018). [Cobalt: A Critical and Strategic Element. *Annals of the National Academy of Exact, Physical and Natural Sciences* 70, 77-106 (2018)].
15. Baran, E. J., B. S. Parajón-Costa & R. C. Mercader: "Spectroscopic Characterization of $NH_4FeP_2O_7$," *J. Mol. Struct.* 1188, 234-237 (2019).
16. León, I. E., M. C. Ruiz, C. A. Franca, B. S. Parajón-Costa & E. J. Baran: "Metvan, bis(4,7-Dimethyl-1,10-phenanthroline) sulfatooxidovanadium(IV): DFT and Spectroscopic Study – Antitumor Action on Human Bone and Colorectal Cancer Cell Lines," *Biol. Trace Elem. Res.* 191, 81-87 (2019).
17. Cacicedo, M. L., M. C. Ruiz, S. Scioli-Montoto, M. E. Ruiz, M. A. Fernández, R. M. Torres-Sánchez, E. J. Baran, G. R. Castro & I. E., León: "Lipid Nanoparticles-Metvan: Revealing a Novel Way to Deliver a Vanadium Compound to Bone Cancer Cells," *New J. Chem.* 43, 17726-17734 (2019).
18. González-Baró, A. C., V. L. Barone & E. J. Baran: "Vibrational Spectra of Two Bismuth (III) Oxalates: $Bi(OH)C_2O_4$ and $Bi_2(C_2O_4)_3 \cdot 7H_2O$," *Anales Asoc. Quím. Argent.*, in the press.
19. Balsa, L. M., M. C. Ruiz, L. Santamaría de la Parra, E. J. Baran & I. E., León: " Anticancer and antimetastatic activity of copper(II) tropolone complex against human breast cancer cells, breast multicelular spheroids and mammospheres," *J. Inorg. Biochem.*, in the press.

Books and book chapters:
1. Baran, E. J. (Editor): "Litio. Un Recurso Natural Estratégico," Academia Nacional de Ciencias Exactas, Físicas y Naturales, Buenos Aires (2017). ISBN:978-987-4111- 22-7, iv + 240 pp. [Lithium. A Natural Strategic Resource. National Academy of Exact, Physical and Natural Sciences, Buenos Aires, 2017].
2. Baran, E. J.: "Nickel and its Role in Plant Physiology" en *Advances in Plant Physiology* (H. Hemantaranjan, Ed.), Scientific Publishers, Jodhpur, Vol. 17, 291-313 (2017).
3. Baran, E. J.: "Plant Purple Acid Phosphatases: Structure and Functions" en *Advances in Plant Physiology* (H. Hemantaranjan, Ed.), Scientific Publishers, Jodhpur, Vol. 17, 331-359 (2017).
4. Baran, E. J.: "Lithium in Plants" in *Advances in Plant Physiology* (H. Hemantaranjan, Ed.), Scientific Publishers, Jodhpur, Vol. 18, 155-162 (2019).

In: Oxalate
Editor: Elsa Kytönen

ISBN: 978-1-53618-303-0
© 2020 Nova Science Publishers, Inc.

Chapter 4

THE ASSOCIATION OF DIABETES AND OBESITY WITH UROLITHIASIS: A REVIEW

*Mohd Sajad and Sonu Chand Thakur**
Centre for Interdisciplinary Research in Basic Sciences,
Jamia Millia Islamia, New Delhi, India

ABSTRACT

Hyperoxaluria can induce kidney disease by several pathways, including calcium oxalate crystal tubular blocking, inflammation, and injury in epithelial tubular cell. Hyperoxaluria is also seen in people with obesity and mellitus, which are extenuating features that lead to chronic kidney disease (CKD). In addition. This is not still clear that hyperoxaluria is a possible cause behind this increased incidence of CKD in diabetes mellitus and obesity. These are linked with higher excretion of urinary oxalate through a variety of pathways. Hyperoxaluria can be a pathway through which the kidney dysfunction occurs in persons due to diabetes mellitus or obesity and thus, lead to a gradual deterioration of renal function. Potential work on pharmacological or dietary

---

*Corresponding Author's Email: sthakur@jmi.ac.in.

interventions to minimize the production or absorption of oxalate is required to check whether the reducing excretion of urinary oxalate is helpful in inhibiting the development of kidney disease and the process of obesity and diabetes mellitus.

Keywords: diabetes, hyperoxaluria, metabolic syndrome, chronic kidney disease, oxalate nephropathy

1. INTRODUCTION

Both obesity and diabetes mellitus are a significant epidemiological problem around the world particularly in the United States. Both disorders lead to early death then increased morbidity, particularly kidney failure. Diabetes is extreme cause of chronic kidney disease (CKD) and also the end-stage renal disease (ESRD) worldwide, and obesity is probably lead to the incidence and development of CKD. These two conditions (diabetes and obesity) also lead to nephrolithiasis. The most common constituent of kidney stones is Calcium oxalate [1]. Ironically, the main risk factors for the excretion of higher urinary oxalate are both obesity and diabetes. The secretion of oxalate could increase in these two cases and thus lead to an elevated probability of nephrolithiasis[2, 3]. Currently, animal experiments and a potential cohort studies have been shown a correlation between oxalates and the production and progression of CKD [4, 5]. Primary hyperoxaluria is an unusual inborn defect in the synthesis of glyoxylate caused by overproduction of oxalate which is deposited as calcium oxalate in different organs as renal function declines. The kidney is the key focus for oxalate deposition, contributing to oxalate nephropathy and renal failure. Secondary hyperoxaluria is more frequent and is typically the consequence of elevated di-oxalate deposition [6]. Oxalate, is the main component in 70-80 percent of kidney stones plays a crucial function in the forming of stones. Calcium oxalate is more supersaturated in stone-forming urine than in safe subjects6 and oxalate concentrations 25 times higher in stone-forming papilla than in urine have been recorded, leading to the development of calcium deposits[7]. The evidence that hyperoxaluria can involve as a different mediator in the pathology of CKD and also in diabetes and obesity advances is presented in this chapter.

2. OVERVIEW OF OXALATE METABOLISM

The terminal metabolite in humans is oxalate ($C_2O_4^{2-}$-dicarboxylic acid). Oxalate is derived in blood and urine both exogenously and endogenously from diet by metabolism[8] Fig. 1. The typical pathways of glyoxylate metabolism converge on the instant precursor of oxalate and recently have been recognized as a possible metabolite predictor of type 2 diabetes mellitus [9]. Endogenous synthesis primarily occurs in liver. Amino acid (serine, hydroxyproline, tryptophan, glycine), ascorbic acid, fructose, glucose, and glycol metabolism produce glyoxylate *in vivo*[10]. Glyoxal, which is a result of cellular peroxidization and protein glycation, can also be used as a source to oxalate. The glyoxal can either directly or indirectly be converted into oxalate [11]. Several foods comprise of oxalate, that is absorbed in stomach and mostly in the small and large intestine [12]. Spinach, nuts, potatoes and chocolate are foods that contain high levels of oxalate. Average consumption and absorption of daily oxalate vary among people as interpersonal absorption variations and geographical variations in the oxalate quality of food are different. The presence of calcium free in the intestine is a significant determinant of oxalate absorption: calcium is efficiently and quickly absorbed with oxalates as a CaOx complex, which are even less absorbed than free form of oxalate [13]. Passive and transcellular absorption of the intestinal oxalate is conducted with differing levels of absorptions vs. secretions in the gastrointestinal tract [14, 15]. Transcellular transportation is mediated by the anion exchanger SLC26 which is expressed on apical and basolateral membranes of the human colon and small gut. Research indicates that inflammatory changes in the transport of bowel oxalate may account for hyperoxaluria caused by obesity, leading to supprimation or decreased absorption of active oxalate from the bowel [15]. The degradation of oxalate occur in colon by metabolizing oxalate in gut along with the formigenes of Oxalobacter and other bacteria [16, 17]. Faecal discharges account for less than 10% of oxalate discharges. The excretion of oxalate is easily filtered, absorbed and isolated by the proximal tubular [18] in the glomerulus. The 24-hour excretion of oxalate in urine is a result of intakes of oxalate by diet, absorption by intestinal and endogenous hepatitis oxalate synthesis. And hyperoxaluria can be caused by many factors. Clinical factors related to higher excretion of urinary oxalate are age, diabetes, higher BMI, and decreased oxalate and fructose usage in human needs that analyzed in 24-hour excretion of urinary oxalate [19]. In the research, CKD persons featured overweight, white vs. black skin, diuretic thiazide, lower excretion of urinary calcium and low level of calcium in serum [4], specific variables correlated with high excretion of urinary oxalate. Hyperoxaluria also occurs in primary hyperoxalurias, autosomal recessive conditions, owing to the decreased intestinal accumulation of oxalate caused by enzymes mutations which are involved in metabolism of oxalate and enteric hyperoxaluria.

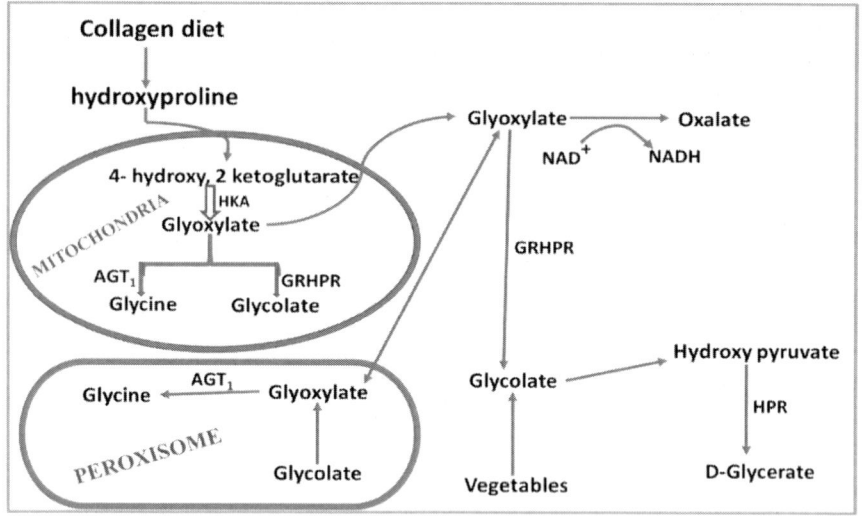

Figure 1. Depicting the mechanism of Oxalate metabolism.

Table 1. Stone, Urinary risk factors and Treatment

Risk factors	Causes	Treatment
Low Urine	Exercise, Sweating, Low fluid intake, Bowel disease	Scheduled fluid intake
Hypercalciuria	Idiopathic, Hyperparathyroidism, Distal renal Tubular acidosis	Reduce sodium and protein intake
Hyperoxaluria	High oxalate, Low Calcium diet, Inherited, Inflammatory bowel disease	Low oxalate diet, adequate calcium intake, calcium and magnesium intake
Hypocitraturia	Idiopathic, high protein diet, Potassium depletion	Reduce proteinaceous diet
Hyperuricosuria	High purine diet	Reduce purine diet

3. EXCRETION OF URINARY OXALATE AND DIABETES MELLITUS

Across several trials, the excretion of urine with oxalate across people with vs. diabetes mellitus was shown to be greater. In kidney stones the calcium oxalate stones are the typical form [20] and Diabetes mellitus is related by an elevated risk of developing kidney stones. Diabetes mellitus has been independently linked with higher excretion of urinary oxalate in 3123 individuals with proven CKD: and 24-hour excretion of urinary oxalate in those individuals with diabetes mellitus had a 11% higher than those individuals without diabetes following changes to a variety of factors including drugs, age, BMI, sex, race, and also the laboratory tests [4]. Through metabolomic profiling of plasma in humans, glyoxylate, the precursor immediate to oxalate has been identified as the possible diabetes mellitus metabolite marker [9, 21]. A long-term blood donation analysis shows that higher plasma levels were diagnosed with diabetes mellitus up to 3 years earlier in age, sex and BMI analyses [9]. The levels of Glyoxylate were six times much in diabetic mice than in control mice [9]. Glyoxal, an oxodehyde, which can be produced by protein glycation or lipid peroxidation by diabetes from hyperglycaemia[22], is another possible precursor of oxalate. Glyoxal was thought to be an important human use and cause of oxidative stress for endogenous oxalate synthesis [11, 23]. Glyoxal, which is an a-oxodehyde, produced by glycation of protein or lipid peroxidation during hyperglycaemia [28,30], is another possible precursor of oxalate. Glyoxal was thought to be an important human use and cause of oxidative stress for endogenous oxalate synthesis [11, 23]. In subsequent trials in 1003 Type II, and 159 Type 1 diabetes patients, a-oxoaldehydes and methylglyoxal also linked with cardiovascular disease (CDI). In 1481 screen-detected type 2 diabetic patients, the baseline and 6-year-longitudinal level of methylglyoxal was reversely associated with projected glomerular filtration levels (eGFR). A retrospective clinical analysis of more than 150 persons with CKD phases of 3–5 correlated elevated methyl-levels with an improved likelihood of progression to ESRD more than two years and more than six years respectively [24].

4. OBESITY AND URINARY OXALATE EXCRETION

In Medical Professionals Follow-Up Survey, Nurses Health Research, and Nurses Health Report II, Lower BMI was separately related to greater urinary oxalate excretion [2]. In the Chronic Renal Insufficiency Cohort (CRIC) analysis, higher BMI in unadjusted analyses was correlated with higher excretion of urinary oxalate but not after multivariable changes [4]. Obesity is regarded as a contributing factor for nephrolithiasis [25], most often linked with calcium oxalate forming stones. Two recent research recognized the hyperoxaluria pathways and emphasizing the role of inflammation in obesity [26, 27]. Amin et al, find data to support decreased the secretion of active oxalate in intestinal from local and systemic inflammation as a trigger for a decrease in faecal oxalate excretion. Bashir et al. have found signs of decreased paracellular oxalate absorption all over the gastrointestinal tract. The elevated oxidative stress and proinflammatory cytokines in obesity, greatly decreased the *in vitro* and *ex vivo* oxalate absorption from paracellular intestines. Such findings demonstrated that inflammation significantly increased the gut absorption and reduced gut secretion as a consequence of obesity-associated hyperoxaluria.

5. EVIDENCE ON THE PROGRESSION BETWEEN URINARY OXALATE EXCRETION AND CHRONIC KIDNEY DISEASE

In 3123 CRIC participants [4], increased level in excretion of urinary oxalate were observed as a risk factor for the development of kidney disease thus, lead to ESRD. Higher excretion of urinary oxalate was found in less eGFR and high albuminuria patients cross-sectionally. The highest signs were found in people with lost kidney function due to higher BMI (45 percent higher risk of ESRD) and with diabetes mellitus (44 percent higher risk of ESRD). A comprehensive analysis of reported cases of secondary oxalate nephropathy verified by biopsy identified 108 patients with oxalate nephropathy, of which none had full recovery, 42 per cent had partial recovery, and 58 per cent had no recovery and remained reliant on dialysis [6]. Several animal studies [28, 29] have also investigated the processes by which oxalate can induce kidney damage. As a central mechanism underlying the discovery, sterile inflammation of the intracellular nucleotide-binding network, leucine rich repeat-containing receptor, pyrin-containing-3 (NLRP3) inflammatory activation was examined [29]. CKD models of oxalate feed have also been developed as a reproducible CKD protocol that recapitulates clinical manifestations of CKD in human [30]. Recently, Saenz et al. observed that metabolic syndrome leads to hyperoxaluria-induced renal damage in the murine nephrolithiasis model, consistent with the CRIC finding of a greater oxalate signal correlated with renal impairment in those with a higher BMI.

6. PATHOPHYSIOLOGY

The association between nephrolithiasis and obesity has been significantly studied in last three decades and the results have demonstrated that apparent variations in activity, in diet, acid / base imbalance in renal, and other alterations in urinary chemistry [31]. It is evident that the large variation in dietary obesity is mostly correlated with a rise in caloric consumption, which has been shown to present a danger for the development of urinary stones [32]. Supporting this theory, in a report based on variations in eating patterns of postmenopausal people, higher calorie consumption reduced physical activity and increased BMI was correlated with urolithiasis [33]. With respect to the potential underlying reasons for a critical shift in pH of urine to acidic level, secretion of insulin in obesity and decreased development of fatty acids (FAs) may be main causes for these modifications. While several experiments have indicated increases in the expression of insulin receptors in renal tubular epithelium cells thus, resulting in an increase in hydrogen ions in the urine [34, 35], the higher insulin condition, along with a rise in FAs, may result in a reduction in renal ammoniagenesis [36]. Due to these changes, a reduced urinary buffering ability may also result in lowering the pH of urine, which would raise the likelihood of calcium oxalate and uric acid stone formation. In addition to urinary pH variations, some research centred on potential alterations in the content of urine in these situations with regard to the secretion of those risk factors that lead to the formation of urinary stone. The findings of a comprehensive review, involving 6000 participants, found that decreased excretion of uric acid, urinary sodium, phosphate and sulphate rates possibly shown in patients with obesity relative to average cases [37]. In addition, decreased excretion of uric acid, oxalate, sodium, cystine, and sulphate was also studied in overweight patients [38] in which these patients excreted more uric acid, sodium, and calcium in urine as compared to normal [3]. Further highlighting the increased incidence of urinary disorders, Ekeruo et al. demonstrated that overweight patients with urinary stone had severe hyperuricosuria and hypocitraturia (43 and 54%) relative to non-obese urolithiasis equivalents

[39]. Finally, through re-studying the variations and risk factors during the process of formation of urinary stone, Taylor and Curhan and Powell et al. also showed the elevated excretion rates of uric acid, oxalate, sodium, phosphate, cysteine and sulphate in urine in obese and non-obese patients [40, 41]. As these results specifically shows that the urinary content of obese patients tends to include elevated rates of those conditions that results in stone-forming relative to those of general population. Among the potential underlying pathophysiological pathways suggested so far, Canada and Isgoren, in their highly original research, argued that the reduced activity of neural tissue and interstitial cells within the urothelial integrity of diabetic rabbits that reduce ureteral peristalsis thus resulting in the urinary stasis and formation of stone [42]. In diabetic cases, however, the mechanism of insulin resistance has been responsible for stone forming. When shown by the report, insulin resistance lead to acidic urine status in those situations which arising from compromised renal ammoniagenesis. It also improves the reabsorption of uric acid in renal proximal tubules, resulting in proof of hyperuricemia in diabetic situations. Thus, each of these factors may be responsible for the most often found stone composition 'uric acid' in obese cases of diabetes [43]. Finally, research have shown that hyperglycaemia can similarly correlated with decreased level of oxalate and calcium in urinary excretion [44].

Table 2. Anti-urolithiatic activity of some medicinal plants

S. NO	Medicinal Plants	Mechanism of Action
1	*Ammivisnaga*	Antilithiatic and diuretic[49]
2	*Phoenixdactyleferae*	Antiurolithiatic[50]
3	*Phyllanthusniruri*	Inhibitory effect on crystal growth[51]
4	*Moringaoleifera*	Reduced urinary oxalate, regulatory action on endogenous oxalate synthesis in hyperoxaluria [52]
5	*Benincasahispida*	Antiurolithiatic[53]
6	*Acoruscalamus*	Diuretic, strongly suppressing various urolithiatic promotors[54]
7	*Aervajavanica*	Preventing growth of urinary stones, supporting folk information[55]

Table 2. (Continued)

S.NO	Medicinal Plants	Mechanism of Action
8	*Craetavanurvala*	Preventing the deposition of calcium and oxalate[55]
9	*Aervalanata*	Antiurolithiatic[17], Antiurolithiatic & diuretic[56]
10	*Asparagusracemosus*	Antiurolithiatic[57]
11	*Biophytumsensitivum*	Diuretic, Antiurolithiatic[58]
12	*Ipomoeaeriocarpa*	Inhibits the growth of urinary stones [59]
13	*Meliaazedarach*	Reduced urinary calcium, oxalate, phosphate [60]
14	*Daucuscarota*	Anticrystallization, reduction in crystalsize, inhibitory effect on CaOx crystal aggregation[61]
15	*Ichnocarpusfrutescens*	Reducing the risk of CaOx supersaturation[62]
16	*Viburnumopulus*	Inhibition of oxalate,free radical production & Diuretic[63]
17	*Aervajavanica*	Preventing growth of urinary stones, supporting folk information[55]
18	*Tripalakarpachooranam*	CaOx crystal inhibitory, Diuretic, Epithelial cell protective, Hypocalciuric and Hypercitrauric effects[64]
19	*Peltophorumpterocarpum*	More effective in dissolving calcium oxalate[65]
20	*Nothosaervabrachiate*	Inhibition of oxalate, free radical production & Diuretic[63]

7. OXIDATIVE STRESS AND UROLITHIASIS

Epidemiological trials have established proof of interaction among nephrolithiasis and a variety of cardiovascular disorders, together with obesity, chronic kidney disease diabetes, and metabolic syndrome. Several co-morbidities that not only contribute to stone disease, but may also be caused by it. Nephrolithiasis is a contributing factor for developing hypertension and has a greater incidence of diabetes mellitus, although certain hypertensive although diabetic patients are at increased risk for stone forming[45]. An overview of the relationship between stone disease and other associated disorders, as well as the factors involved in their pathogenesis, that provide insight into the development of stone. It is our belief that the correlation between the shape of stones and the occurrence of co-morbidities is the product of some common pathological characteristics. Recent literature reviews suggest that the growth of reactive oxygen species (ROS) and the formation of oxidative stress (OS) could be such a typical pathway[46]. OS is a popular characteristic in any cardiovascular disease (CVD) including diabetes mellitus, hypertension, myocardial infarction, and atherosclerosis. There is growing indication that ROS is also developed during idiopathic calcium oxalate (CaOx) nephrolithiasis.Both tissue culture and animal model experiments have shown that ROS is formed during association between Calcium Phosphate (CaP) crystals/CaOx and renal epithelial cells. Clinical trials have also provided proof of the production of oxidative stress in stone-forming patients 'kidneys. Renal conditions progressing to OS tend to be a spectrum. Stress triggered by one condition can activate the other under the right circumstances[47].

CONCLUSION

In both developed and emerging countries, the incidence of kidney stones and obesity is increasingly growing. Patients with obesity disorders have specific risk factors for urolithiasis relative to their usual- peers. Insulin tolerance, acid-base balance disorders, variations in urinary composition, and related eating patterns complications are the major underlying factors that raise the likelihood of urolithiasis. Such cases required to undertake an additional rigorous metabolic analysis to determine the underlying serum and urinary variations in order to develop an appropriate treatment plan. Current metabolic assessment-based targeted medicinal treatment and well-planned nutritional interventions can be helpful in the prevention of stone forming and in the correction of urinary irregularities. Weight reduction under close monitoring is a valuable alternative indicator for a positive result. Patient with obesity disorders have specific risk factors for urolithiasis relative to non-obese peers. Any effort must be applied to raise knowledge of this issue and the required steps and a well-planned recovery program are the key tasks for urologists.

REFERENCES

[1] Jha, V., et al., Chronic kidney disease: global dimension and perspectives. *The Lancet*, 2013. 382(9888): p. 260-272.

[2] Taylor, E.N. and G.C. Curhan, Determinants of 24-hour urinary oxalate excretion. *Clinical Journal of the American Society of Nephrology*, 2008. 3(5): p. 1453-1460.

[3] Taylor, E.N. and G.C. Curhan, Body size and 24-hour urine composition. *American journal of kidney diseases*, 2006. 48(6): p. 905-915.

[4] Waikar, S.S., et al., Association of urinary oxalate excretion with the risk of chronic kidney disease progression. *JAMA internal medicine*, 2019. 179(4): p. 542-551.

[5] Knauf, F., et al., NALP3-mediated inflammation is a principal cause of progressive renal failure in oxalate nephropathy. *Kidney international*, 2013. 84(5): p. 895-901.

[6] Lumlertgul, N., et al., Secondary oxalate nephropathy: a systematic review. *Kidney international reports*, 2018. 3(6): p. 1363-1372.

[7] Batagello, C.A., M. Monga, and A.W. Miller, Calcium oxalate urolithiasis: a case of missing microbes? *Journal of endourology*, 2018. 32(11): p. 995-1005.

[8] Holmes, R.P., H.O. Goodman, and D.G. Assimos, Contribution of dietary oxalate to urinary oxalate excretion. *Kidney international*, 2001. 59(1): p. 270-276.

[9] Nikiforova, V.J., et al., Glyoxylate, a new marker metabolite of type 2 diabetes. *Journal of diabetes research*, 2014. 2014.

[10] Knight, J., et al., Metabolism of fructose to oxalate and glycolate. *Hormone and Metabolic Research*, 2010. 42(12): p. 868-873.

[11] Lange, J.N., et al., Glyoxal formation and its role in endogenous oxalate synthesis. *Advances in urology*, 2012. 2012.

[12] Hatch, M. and R.W. Freel, Intestinal transport of an obdurate anion: oxalate. *Urological research*, 2005. 33(1): p. 1-16.

[13] von Unruh, G.E., et al., Dependence of oxalate absorption on the daily calcium intake. *Journal of the American Society of Nephrology*, 2004. 15(6): p. 1567-1573.

[14] Whittamore, J.M. and M. Hatch, The role of intestinal oxalate transport in hyperoxaluria and the formation of kidney stones in animals and man. *Urolithiasis*, 2017. 45(1): p. 89-108.

[15] Sakhaee, K., Unraveling the mechanisms of obesity-induced hyperoxaluria. *Kidney international*, 2018. 93(5): p. 1038-1040.

[16] Liebman, M. and I.A. Al-Wahsh, Probiotics and other key determinants of dietary oxalate absorption. *Advances in Nutrition*, 2011. 2(3): p. 254-260.

[17] Hatch, M., et al., Oxalobacter sp. reduces urinary oxalate excretion by promoting enteric oxalate secretion. *Kidney international*, 2006. 69(4): p. 691-698.

[18] Hatch, M. and R.W. Freel, Renal and intestinal handling of oxalate following oxalate loading in rats. *American journal of nephrology*, 2003. 23(1): p. 18-26.

[19] Otto, B.J., et al., Age, body mass index, and gender predict 24-hour urine parameters in recurrent idiopathic calcium oxalate stone formers. *Journal of endourology*, 2017. 31(12): p. 1335-1341.

[20] Chung, S.-D., Y.-K. Chen, and H.-C. Lin, Increased risk of diabetes in patients with urinary calculi: a 5-year followup study. *The Journal of urology*, 2011. 186(5): p. 1888-1893.

[21] Padberg, I., et al., A new metabolomic signature in type-2 diabetes mellitus and its pathophysiology. *PloS one*, 2014. 9(1): p. e85082.

[22] Lapolla, A., et al., Glyoxal and methylglyoxal levels in diabetic patients: quantitative determination by a new GC/MS method.

Clinical Chemistry and Laboratory Medicine, 2003. 41(9): p. 1166-1173.

[23] Wang, X.-J., et al., Elevated levels of α-dicarbonyl compounds in the plasma of type II diabetics and their relevance with diabetic nephropathy. *Journal of Chromatography B*, 2019. 1106: p. 19-25.

[24] Tezuka, Y., et al., Methylglyoxal as a prognostic factor in patients with chronic kidney disease. *Nephrology*, 2019. 24(9): p. 943-950.

[25] Carbone, A., et al., *Obesity and kidney stone disease: a systematic review*. 2018.

[26] Bashir, M., et al., Enhanced gastrointestinal passive paracellular permeability contributes to the obesity-associated hyperoxaluria. *American Journal of Physiology-Gastrointestinal and Liver Physiology*, 2019. 316(1): p. G1-G14.

[27] Amin, R., et al., Reduced active transcellular intestinal oxalate secretion contributes to the pathogenesis of obesity-associated hyperoxaluria. *Kidney international*, 2018. 93(5): p. 1098-1107.

[28] Khan, S.R., Crystal-induced inflammation of the kidneys: results from human studies, animal models, and tissue-culture studies. *Journal of Clinical and Experimental Nephrology*, 2004. 8(2): p. 75-88.

[29] Mulay, S.R., et al., Calcium oxalate crystals induce renal inflammation by NLRP3-mediated IL-1β secretion. *The Journal of clinical investigation*, 2012. 123(1).

[30] Mulay, S.R., et al., Oxalate-induced chronic kidney disease with its uremic and cardiovascular complications in C57BL/6 mice. *American Journal of Physiology-Renal Physiology*, 2016. 310(8): p. F785-F795.

[31] Asplin, J.R., Obesity and urolithiasis. *Advances in chronic kidney disease*, 2009. 16(1): p. 11-20.

[32] Powell, C.R., et al., Impact of body weight on urinary electrolytes in urinary stone formers. *Urology*, 2000. 55(6): p. 825-830.

[33] Sorensen, M.D., et al., Activity, energy intake, obesity, and the risk of incident kidney stones in postmenopausal women: a report from

[34] Taylor, E.N., T.T. Fung, and G.C. Curhan, DASH-style diet associates with reduced risk for kidney stones. *Journal of the American Society of Nephrology*, 2009. 20(10): p. 2253-2259.

[35] Fuster, D.G., et al., Characterization of the regulation of renal Na+/H+ exchanger NHE3 by insulin. *American Journal of Physiology-Renal Physiology*, 2007. 292(2): p. F577-F585.

[36] Chobanian, M.C. and M.R. Hammerman, Insulin stimulates ammoniagenesis in canine renal proximal tubular segments. *American Journal of Physiology-Renal Physiology*, 1987. 253(6): p. F1171-F1177.

[37] Bobulescu, I.A., Renal lipid metabolism and lipotoxicity. *Current opinion in nephrology and hypertension*, 2010. 19(4): p. 393.

[38] Taylor, E.N., M.J. Stampfer, and G.C. Curhan, Obesity, weight gain, and the risk of kidney stones. *JAMA*, 2005. 293(4): p. 455-462.

[39] Del Valle, E.E., et al., Metabolic diagnosis in stone formers in relation to body mass index. *Urological research*, 2012. 40(1): p. 47-52.

[40] Ekeruo, W.O., et al., Metabolic risk factors and the impact of medical therapy on the management of nephrolithiasis in obese patients. *The Journal of Urology*, 2004. 172(1): p. 159-163.

[41] Assimos, D.G., Fructose consumption and the risk of kidney stones. Journal of endourology, 2008. 22(5): p. 853-854.

[42] Powell, C., et al., Erratum: Impact of body weight on urinary electrolytes in urinary stone formers (Urology 55: June 2000 (825-830)). *Urology*, 2000. 56(2): p. 352.

[43] Canda, A.E. and A.E. Isgoren, Re: Increased risk of diabetes in patients with urinary calculi: a 5-year followup study: S.-D. Chung, Y.-K. Chen and H.-C. Lin J Urol 2011; 186: 1888-1893. *The Journal of urology*, 2012. 187(6): p. 2279.

[44] Abate, N., et al., The metabolic syndrome and uric acid nephrolithiasis: novel features of renal manifestation of insulin resistance. *Kidney international*, 2004. 65(2): p. 386-392.

[45] Sarica, K., Obesity and stones. *Current opinion in urology*, 2019. 29(1): p. 27-32.

[46] Pearle, M.S., et al., Urologic diseases in America project: urolithiasis. *The Journal of urology*, 2005. 173(3): p. 848-857.

[47] Li, X., et al., Anti-nephrolithic potential of catechin in melamine-related urolithiasis via the inhibition of ROS, apoptosis, phospho-p38, and osteopontin in male Sprague-Dawley rats. *Free radical research*, 2015. 49(10): p. 1249-1258.

[48] Khan, S.R., Is oxidative stress, a link between nephrolithiasis and obesity, hypertension, diabetes, chronic kidney disease, metabolic syndrome? *Urological research*, 2012. 40(2): p. 95-112.

[49] Khan, Z.A., et al., Inhibition of oxalate nephrolithiasis with Ammi visnaga (AI-Khillah). *International urology and nephrology*, 2001. 33(4): p. 605-608.

[50] Al-Gamli, A., et al., Evaluation of anti urolithiatic activity of phoenix dactyleferae seeds extract in ethylene glycol induced urolithiasis in rats. *IJPPR Human*, 2017. 9: p. 6-20.

[51] Freitas, A., N. Schor, and M.A. Boim, The effect of Phyllanthus niruri on urinary inhibitors of calcium oxalate crystallization and other factors associated with renal stone formation. *BJU international*, 2002. 89(9): p. 829-834.

[52] Karadi, R.V., et al., Effect of Moringa oleifera Lam. root-wood on ethylene glycol induced urolithiasis in rats. *Journal of ethnopharmacology*, 2006. 105(1-2): p. 306-311.

[53] Patel, R., S. Patel, and J. Shah, Anti-urolithiatic activity of ethanolic extract of seeds of Benincasa hispida (Thumb). *Pharmacologyonline*, 2011. 3: p. 586-591.

[54] Ghelani, H., M. Chapala, and P. Jadav, Diuretic and antiurolithiatic activities of an ethanolic extract of Acorus calamus L. rhizome in experimental animal models. *Journal of traditional and complementary medicine*, 2016. 6(4): p. 431-436.

[55] Padala, K. and V. Ragini, Anti urolithiatic activity of extracts of Aerva javanica in rats. *International Journal of Drug Development and Research*, 2014. 6(4): p. 35-45.

[56] Dinnimath, B.M., S.S. Jalalpure, and U.K. Patil, Antiurolithiatic activity of natural constituents isolated from Aerva lanata. *Journal of Ayurveda and integrative medicine*, 2017. 8(4): p. 226-232.

[57] Jagannath, N., et al., Study of antiurolithiatic activity of Asparagus racemosus on albino rats. *Indian journal of pharmacology*, 2012. 44(5): p. 576.

[58] Pawar, A.T. and N.S. Vyawahare, Protective effect of standardized extract of Biophytum sensitivum against calcium oxalate urolithiasis in rats. *Bulletin of Faculty of Pharmacy, Cairo University*, 2015. 53(2): p. 161-172.

[59] Das, M. and H. Malipeddi, Antiurolithiatic activity of ethanol leaf extract of Ipomoea eriocarpa against ethylene glycol-induced urolithiasis in male Wistar rats. *Indian journal of pharmacology*, 2016. 48(3): p. 270.

[60] Christina, A., et al., Antilithiatic Effect of Melia azedarach. on Ethylene Glycol–Induced Nephrolithiasis in Rats. *Pharmaceutical biology*, 2006. 44(6): p. 480-485.

[61] Bawari, S., A.N. Sah, and D. Tewari, Antiurolithiatic Activity of Daucus carota: An in vitro Study. *Pharmacognosy Journal*, 2018. 10(5).

[62] Anbu, J., et al., Antiurolithiatic activity of ethyl acetate root extract of Ichnocarpus frutescens using ethylene glycol induced method in rats. *Journal of Pharmaceutical Sciences and Research*, 2011. 3(4): p. 1182.

[63] İlhan, M., et al., Preclinical evaluation of antiurolithiatic activity of Viburnum opulus L. on sodium oxalate-induced urolithiasis rat model. *Evidence-Based Complementary and Alternative Medicine*, 2014. 2014.

[64] Sudha, V., Antiurolithiatic activity of medicinal plants and Siddha Formulatory Medicines: A Review. *Journal of Research in Biomedical Sciences*, 2020. 3(1): p. 07-12.

[65] Jha, R., et al., Phytochemical analysis and in vitro urolithiatic activity of Peltophorum pterocarpum leaves (DC) Baker. *Journal of medicinal plants studies*, 2016. 4(3): p. 18-22.

INDEX

#

2,3-diketogulonic acid, 102

A

amino acid(s), 12, 21, 23, 106, 111
ammonium, 10, 11, 35, 45, 58, 61, 69, 70, 73, 96
analgesic, 13, 14, 32, 37, 40, 41, 42, 43
antioxidant, 3, 13, 14, 31, 33, 34, 35, 38, 39, 40, 42, 46, 47, 53
aromatic compounds, 107
ascorbic acid, vii, ix, 14, 38, 59, 79, 95, 96, 99, 100, 101, 102, 103, 104, 105, 121, 123, 125, 135
autosomal recessive, 11, 136

B

Bacillus subtilis, 110, 111, 119, 123, 124, 125
bacteria, ix, 11, 95, 96, 105, 110, 117, 118, 119, 123, 124, 136

batteries, 73, 75, 76, 78, 80, 84
biosynthesis, vii, ix, 95, 98, 100, 101, 102, 103, 104, 121, 122, 124
blood urea nitrogen, 32, 35
bronchitis, 38
bronchopneumonia, 39

C

calcium, viii, ix, 2, 3, 6, 8, 9, 10, 12, 13, 15, 19, 20, 21, 22, 23, 24, 25, 26, 30, 34, 35, 36, 37, 38, 39, 41, 42, 43, 46, 48, 52, 96, 97, 98, 100, 103, 108, 118, 121, 123, 124, 126, 133, 134, 135, 137, 138, 140, 142, 143
calcium carbonate, 12, 19, 52
calcium channel blocker, 30
calcium oxalate, viii, ix, 2, 3, 6, 10, 12, 13, 20, 21, 22, 23, 25, 26, 35, 38, 41, 42, 43, 48, 49, 50, 52, 96, 97, 98, 100, 103, 108, 119, 121, 122, 123, 124, 126, 133, 134, 137, 138, 140, 142, 143, 145, 148, 149
caoxite, 96

chronic kidney disease (CKD), ix, 2, 11, 133, 134, 136, 137, 139, 142, 144, 146, 148
cobalt, 77, 79, 106
copper, 61, 78, 96, 106, 107, 130
crystal growth, viii, 1, 19, 22, 25, 52, 141
crystal idioblasts, 103, 123
crystal structure, 77, 111, 121
crystalline, 2, 12, 18, 49, 70, 79, 96, 97
crystallization, 19, 21, 22, 24, 25, 31, 41, 48, 126
crystals, 7, 10, 13, 15, 18, 19, 20, 21, 24, 25, 26, 34, 38, 39, 42, 59, 96, 97, 98, 103, 143
cupin superfamily, 106, 108, 111, 122

D

D-arabino-1,4-lactone oxidase, 104
D-arabinose, 104
D-arabinose dehydrogenase, 104
decomposition, 65, 66, 67, 68, 69
degradation, ix, 65, 66, 68, 71, 80, 95, 96, 98, 105, 107, 108, 110, 118, 135
degradation process, ix, 68, 95
dehydroascorbic acid, 102
D-erithroascorbic acid, 103
detoxification, 97, 119
diabetes, v, vii, ix, 2, 39, 133, 134, 136, 137, 139, 141, 142, 144, 145, 147, 148

E

energy, vii, viii, 55, 56, 69, 73, 74, 75, 78, 79, 80, 84, 118
energy density, 78, 79
energy storage, v, viii, 55, 56, 73, 74, 75, 84
enzymes, ix, 5, 8, 16, 95, 96, 101, 104, 105, 106, 107, 110, 112, 113, 114, 115, 116, 117, 118, 125, 136
epithelial cells, 6, 13, 19, 20, 38, 143

epithelium, 20, 26, 140
ethylene glycol, 33, 35, 37, 39, 41, 44, 52, 53

F

formyl-CoA-transferase, 118

G

germin, 107, 108, 110, 120, 121, 123, 126
glycolate, 98, 99, 100, 114, 145
glycolate oxidase, 99, 100
glycosylation, 18, 108
glyoxylate, 23, 98, 99, 100, 134, 135, 137, 144

H

hepatic obstruction, 37
hepatitis, 136
hepatocytes, 18
herbal medicine, 31, 36
hydroxyl groups, 14
hypercalciuria, 8, 10, 23
hyperglycaemia, 137, 141
hyperoxaluria, vii, viii, ix, 2, 23, 24, 133, 134, 135, 137, 138, 139, 141, 145, 146
hyperplasia, 5
hyperuricemia, 23, 141

I

ions, 9, 19, 59, 60, 74, 75, 78, 80, 107, 111, 140
iridium, 121
iron, 40, 63, 98, 106, 118, 120, 122
isocitrate, 99, 100
isocitrate lyase, 99

J

jaundice, 37, 38

K

kidney(s), vii, viii, ix, 1, 2, 3, 5, 6, 7, 8, 10, 12, 13, 14, 15, 16, 17, 18, 20, 21, 22, 23, 24, 25, 27, 33, 34, 35, 37, 38, 39, 41, 42, 43, 48, 49, 52, 133, 134, 137, 139, 143
kidney stones, viii, 1, 2, 3, 6, 37, 49, 134, 137, 143

L

L-ascorbic acid, 59, 99, 100, 101, 102, 103, 104, 121, 123
L-galactose, 101, 102
L-galactose dehydrogenase, 101
L-gulono-1,4-lactone, 100
lipid peroxidation, 13, 38, 43, 137
lithium, 56, 76, 77, 81
lithium ion battery(ies), 56, 76
L-threonic acids, 102

M

magnesium, 6, 7, 10, 11, 21, 34, 36, 39, 43, 96, 118, 137
manganese, 79, 96, 106, 107, 110, 111, 113, 114, 117, 121, 124, 125, 126
metabolic syndrome, 24, 134, 139, 142, 147, 148
metabolism, 10, 21, 23, 31, 102, 121, 123, 125, 135, 136
metal ion, 74, 114
metalloenzymes, 117

N

nephrolithiasis, 3, 13, 15, 23, 52, 134, 138, 139, 140, 142
nephron, 8, 9
nephropathy, 17, 49, 134, 139
nickel, 78, 79, 106
nucleation, viii, 1, 19, 23, 24, 25, 31, 40

O

oxalate decarboxylase, ix, 95, 96, 105, 107, 110, 111, 112, 114, 115, 119, 121, 123, 124, 125
oxalate degradation enzymes, ix, 95, 96
oxalate nephropathy, 134, 139, 144
oxalate oxidase, ix, 95, 96, 105, 107, 108, 109, 110, 112, 113, 115, 116, 117, 120, 121, 123, 124, 126
oxalatrophy, 117
oxalic acid, vii, ix, 56, 57, 58, 59, 62, 63, 80, 95, 96, 97, 98, 99, 100, 102, 103, 104, 105, 113, 117, 121, 123, 125
oxaloacetate, 99, 100
oxaloacetate hydrolase, 99
oxalotrophic, ix, 95, 96, 105, 117, 123, 124
oxalotrophic bacteria, ix, 95, 96, 105, 117, 123
oxalyl-CoA-decarboxylase, 118
oxidation, 62, 68, 80, 98, 99, 101, 102, 103, 104, 105, 106, 107, 113
oxidative stress, 6, 13, 52, 138, 143

P

pathogenesis, 10, 12, 17, 48, 49, 143
phosphate(s), 6, 9, 10, 11, 12, 13, 14, 21, 22, 24, 34, 35, 36, 37, 41, 42, 46, 101, 102, 140, 142

proteins, ix, 3, 12, 14, 15, 16, 17, 48, 95, 106, 107, 122, 125, 126
proximal tubules, 21, 141
Pseudomonas aeruginosa, 11

Q

quercitin dioxygenase, 107

R

reaction mechanism, ix, 95, 113, 117, 120
reactive oxygen, 143
reactivity, 74, 119
receptor(s), 16, 22, 23, 39, 48, 139, 140
renal calculi, 11, 32, 34
renal failure, 2, 134
renal medulla, 15, 16

S

salts, 8, 22, 26, 41, 56, 61, 62, 70, 85, 97
storage, viii, 55, 56, 74, 75, 76, 77, 84, 106, 122, 126
substrate(s), vii, ix, 19, 95, 99, 100, 102, 104, 106, 111, 112, 113, 114, 117
synthetic analogues, vii, viii, 56
synthetic pathways, vii, ix, 95, 96

T

tartaric acid, 102
thermal and magnetic properties, 56
thermal decomposition, 65, 69, 71
thermal degradation, 64, 66, 71, 74
thermal properties, 64, 68
transition metal oxalates, v, vii, viii, 55, 56, 58, 65, 73, 74

U

uric acid, 3, 5, 8, 9, 10, 13, 25, 32, 35, 36, 37, 39, 41, 43, 140
uric acid levels, 39
urinary bladder, 29, 32, 37
urinary tract, viii, 1, 2, 28, 29, 36
urine, 2, 6, 7, 8, 9, 11, 12, 13, 15, 16, 17, 18, 19, 21, 22, 23, 24, 25, 27, 31, 32, 33, 34, 35, 37, 39, 41, 42, 43, 46, 134, 135, 137, 140

V

vitamin A, 37
vitamin C, 121, 124, 126
vitamin D, 8, 22
vitamin E, 14
vitamin K, 20

W

weddellite, 10, 96
Wheeler-Smirnoff-mechanism, 96
whewellite, 10, 96

X

X-ray diffraction (XRD), 58, 66
XRD, 58

Z

zinc, 23, 32, 35, 39, 49, 51, 106